LINEAR SYSTEMS PROPERTIES

A Quick Reference

LINEAR SYSTEMS PROPERTIES

A Quick Reference

Venkatarama Krishnan, Ph.D.
University of Massachusetts Lowell
Department of Electrical Engineering
Center for Advanced Computation and Telecommunications

CRC Press
Boca Raton Boston London New York Washington, D.C.

Library of Congress Cataloging-in-Publication Data

Krishnan, Venkatarama, 1929–
Linear systems properties : (quick reference) / Venkatarama Krishnan.
p. cm. -- (Controls)
Includes bibliographical references and index.
ISBN 0-8493-2291-X (alk. paper)
1. Linear systems. 2. Control theory. I. Title. II. Series.
QA402.3.K734 1998
629.8'32—dc21 97-48753
CIP

International Standard Book Number 0-8493-2291-X
Library of Congress Card Number 97-48753
Printed in the United States of America 1 2 3 4 5 6 7 8 9 0
Printed on acid-free paper

PREFACE

The course 16.509-Linear Systems is a one semester graduate course for entering Master's and Doctoral students in Electrical Engineering at the University of Massachusetts Lowell. During discussions in the class, the students felt that it would be helpful if the various formulae and the more important derivations could be given to them as handouts. I decided to produce these handouts in book form. I have updated the material every year for the past 9 years and this book is the updated version, now with added material from Orthogonal functions, Linear Algebra, Singular Value Decomposition and State Space Techniques. Some of the results, in particular the results on complex convolution are novel. I have given a larger number of orthogonal functions than is usually available in handbooks. All results have been tested for accuracy. I hope both graduate and undergraduate students find this book useful as a quick reference for the various formulae encountered in not only Linear Systems but also in other subjects like, Communication Engineering, Control Systems, Signal Processing, Probability, Physics, Electromagnetic Field Theory, etc.

I was able to revise this edition thoroughly and add extra material while on a sabbatical from the University of Massachusetts Lowell during Spring 1997. I am very thankful to the administration for granting me this sabbatical without which it would have been impossible to accomplish this objective. I am also grateful to the Department of Electrical Engineering for recommending this sabbatical and providing me with the facilities. The inputs provided by my graduate students have been invaluable and I appreciate their efforts.

The enthusiastic and cheerful support given to me by associate editor, Nora Konopka, and the excellent proofreading by Mimi Williams made the tedious job of preparing the camera-ready masters a very pleasant one. Finally, I thank the editorial and production staff of CRC Press for the unstinting support they have given me in this effort.

I will greatly appreciate if readers give suggestions for improvement and bring to my attention any errors that they may find.

Lowell,
Massachusetts
Fall 97

विद्या नाम नरस्य रूपमधिकँ प्रच्छन्नगुप्तँ धनँ
विद्या भोगकरी यशःसुखकरी विद्या गुरूणामँ गुरुः।
विद्या बन्धुजनो विदेशगमने विद्या परा देवता
विद्या राजसु पूज्यते नहि धनँ विद्याविहीनः पशुः।।

Maharaja Bhartruhari
Good Sayings

வாக்குண்டாம் நல்ல மனமுண்டாம்
மாமலராள் நோக்குண்டாம் மேனி நுடங்காது
பூக்கொண்டு துப்பார்த் திருமேனி தும்பிக்கையான்
பாதம் தப்பாமல் சார்வார் தமக்கு

*This book is dedicated to
my graduate students,
past and present
who have made
teaching a joy*

CONTENTS

1. MATHEMATICAL FORMULAE

Trigonometric Functions

1. $e^{\pm jx} = \cos x \pm j \sin x$
2. $\cos x = \frac{1}{2}(e^{jx} + e^{-jx})$
3. $\sin x = \frac{1}{2j}(e^{jx} - e^{-jx})$
4. $\cos^2 x + \sin^2 x = 1$
5. $\cos^2 x - \sin^2 x = \cos 2x$
6. $\cos^2 x = \frac{1}{2}(1 + \cos 2x)$
7. $\sin^2 x = \frac{1}{2}(1 - \cos 2x)$
8. $\cos^3 x = \frac{1}{4}(3\cos x + \cos 3x)$
9. $\sin^3 x = \frac{1}{4}(3\sin x - \sin 3x)$
10. $\sin(A \pm B) = \sin A \cos B \pm \cos A \sin B$
11. $\mathrm{Sa}(x) = \frac{\sin x}{x}$: $\mathrm{sinc}(x) = \frac{\sin x}{\pi x}$
12. $\cos(A \pm B) = \cos A \cos B \mp \sin A \sin B$
13. $\tan(A \pm B) = \frac{\tan A \pm \tan B}{1 \mp \tan A \tan B}$
14. $\sin A \sin B = \frac{1}{2}[\cos(A-B) - \cos(A+B)]$
15. $\cos A \cos B = \frac{1}{2}[\cos(A-B) + \cos(A+B)]$
16. $\sin A \cos B = \frac{1}{2}[\sin(A-B) + \sin(A+B)]$

17. $\sin C + \sin D = 2\sin\frac{C+D}{2}\cos\frac{C-D}{2}$

18. $\cos C + \cos D = 2\cos\frac{C+D}{2}\cos\frac{C-D}{2}$

19. $\cos C - \cos D = 2\sin\frac{C+D}{2}\sin\frac{D-C}{2}$

20. $A\cos x + B\sin x = \sqrt{A^2+B^2}\cos(x-\tan^{-1}\frac{B}{A})$

Hyperbolic Functions

21. $\sinh x = \frac{1}{2}(e^x - e^{-x})$

22. $\cosh x = \frac{1}{2}(e^x + e^{-x})$

23. $\cosh x + \sinh x = e^x$

24. $\cosh x - \sinh x = e^{-x}$

25. $e^{\pm x} = \cosh x \pm \sinh x$

26. $\cosh^2 x - \sinh^2 x = 1$

27. $\cosh^2 x + \sinh^2 x = \cosh 2x$

28. $\tanh^2 x + \mathrm{sech}^2 x = 1$

29. $\sinh(A \pm B) = \sinh A \cosh B \pm \cosh A \sinh B$

30. $\cosh(A \pm B) = \cosh A \cosh B \pm \sinh A \sinh B$

31. $\tanh(A \pm B) = \frac{\tanh A \pm \tanh B}{1 \pm \tanh A \tanh B}$

32. $\sinh^2 x = \frac{1}{2}(\cosh 2x - 1)$

33. $\cosh^2 x = \frac{1}{2}(\cosh 2x + 1)$

Exponential Functions

34. $\frac{e^x}{e^y} = e^{(x-y)}$

35. $(e^x)^a = e^{ax}$

36. ln x:– $\ln_e x$: log x:– $\log_i x$, i - integer

37. $\ln xy = \ln x + \ln y$

38. $\ln \frac{x}{y} = \ln x - \ln y$

39. $\ln x^a = a \ln x$

40. $e^{\ln x} = x$: $\ln e = 1$

41. $\log_a x = \log_b x \times \log_a b$, a and b are bases

42. $\ln_e x$, units:- nats : $\log_2 x$, units:- bits

43. # nats = # bits × $\ln_e 2$ = # bits × 0.69315

44. # bits = # nats × $\frac{1}{\ln_e 2}$ = # nats × 1.4427

Power Series

45. $n! = n(n-1)(n-2) \ldots 3.2.1$

46. $e^x = 1 + x + \frac{x^2}{2!} + \ldots + \frac{x^n}{n!} + \ldots$

47. $\ln(1+x) = x - \frac{x^2}{2} + \frac{x^3}{3} + \ldots + (-1)^{(n+1)} \frac{x^n}{n} + \ldots$

48. $(1+x)^n = 1 + nx + \frac{n(n-1)}{2!} x^2 + \ldots$

49. $\frac{1}{(1-x)^n} = 1 + nx + \frac{n(n+1)}{2!} x^2 + \ldots$

2. IMPULSE FUNCTION MODELING

The familiar *loose* definition of the impulse function (Dirac delta function), $\delta(t)$ is:

$$\int_{-\infty}^{\infty} \delta(t - t_0)\, dt = 1 \quad \text{with } \delta(t) = 0 \,,\ t \neq t_0 \qquad (1)$$

Another *loose* definition using the sifting property of the impulse is:

$$\int_{-\infty}^{\infty} \delta(t - t_0)\, x(t)\, dt = x(t_0) \qquad (2)$$

for any $x(t)$ continuous at $t = t_0$. The above integral is not defined in the strict mathematical sense (Riemann or Lebesgue) since it is not of bounded variation. Instead of the above loose definitions, it can be defined more strictly from a distribution (generalized functions) point of view as follows. If $x(t)$ is a continuous function of t in the interval of definition [a, b], and if,

$$\int_a^b x\,(t)\, dt = 1 \quad \textit{(normalization)}$$

then we can write for $a < t_0 \leq b$,

$$\lim_{\varepsilon \to 0} \frac{1}{\varepsilon}\, x\left(\frac{t - t_0}{\varepsilon}\right) = \delta\,(t - t_0) \qquad (3)$$

provided the limiting operation is interpreted in the following distribution (generalized) sense:

$$\lim_{\varepsilon \to 0} \frac{1}{\varepsilon} \int_{-\infty}^{\infty} x\left(\frac{t - t_0}{\varepsilon}\right) g\,(t)\, dt = g\,(t_0) \qquad (4)$$

where $g(t)$ belongs to a set of well-behaved test functions of rapid decay (continuous and not of exponential order). By substituting $\beta = \frac{1}{\varepsilon}$ in eq.(3), the impulse function can also be defined as,

$$\lim_{\beta \to \infty} \beta\, x[\beta(t - t_0)] = \delta(t - t_0) \qquad (5)$$

where again the limiting operation is interpreted in the following sense:

$$\lim_{\beta \to \infty} \beta \int_{-\infty}^{\infty} x[\beta(t - t_0)]\, g(t)\, dt = g(t_0) \qquad (6)$$

Examples

1. $x(t) = \dfrac{1}{\pi\left(1 + t^2\right)}$

$$\delta(t - t_0) = \lim_{\varepsilon \to 0} \frac{\varepsilon}{\pi\left[\varepsilon^2 + (t - t_0)^2\right]}$$

$$= \lim_{\beta \to \infty} \frac{\beta}{\pi\left[1 + \beta^2 (t - t_0)^2\right]}$$

2. $x(t) = \dfrac{1}{2} e^{-|t|}$

$$\delta(t - t_0) = \lim_{\varepsilon \to 0} \frac{1}{2\varepsilon} e^{-\left|\frac{(t - t_0)}{\varepsilon}\right|}$$

$$= \lim_{\beta \to \infty} \frac{\beta}{2} e^{-|\beta(t - t_0)|}$$

3. $x(t) = \dfrac{1}{\sqrt{2\pi}} e^{-t^2/2}$

$$\delta(t - t_0) = \lim_{\varepsilon \to 0} \frac{1}{\varepsilon\sqrt{2\pi}} e^{-(t - t_0)^2/2\varepsilon^2}$$

$$= \lim_{\beta \to \infty} \frac{\beta}{\sqrt{2\pi}} e^{-[\beta(t - t_0)]^2/2}$$

Note that in all the above cases $\int_{-\infty}^{\infty} x(t)\, dt = 1$

3. SIGNAL PROPERTIES

Energy Signals

If x(t) is a real valued signal then the instantaneous power associated with the signal is given by $x^2(t)$. Then the signal energy in a time interval 2T is defined as:

$$E_{2T} = \int_{-T}^{T} | x(t) |^2 dt \qquad (1)$$

The total energy E over the time interval t ε (−∞, ∞) is then written as:

$$E = \lim_{T \to \infty} \int_{-T}^{T} | x(t) |^2 dt \qquad (2)$$

When E is finite then the signal x(t) is called an *energy signal.*

Power Signal

The average power of a signal in the interval 2T can be defined from the instantaneous power as:

$$P_{2T} = \frac{1}{2T} \int_{-T}^{T} | x(t) |^2 dt = \frac{E_{2T}}{2T} \qquad (3)$$

Clearly, if E_{2T} is finite then P_{2T} is infinite. Thus, energy signals have infinite average power. The total average power of a signal is defined by:

$$P = \lim_{T \to \infty} \left[\frac{1}{2T} \int_{-T}^{T} |x(t)|^2 dt \right] = \frac{E}{2T} \qquad (4)$$

If P is finite then x(t) is called a *power signal*. It is clear from the above equation that if P is to be finite then E has to be infinite in such a way that the limit exists. Thus, power signals have infinite energy. Similarly, for energy signals for which E is finite, power P is zero.

Examples
Energy Signal

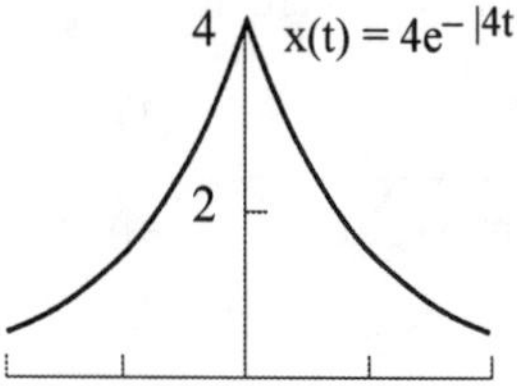

$$\int_{-T}^{T} 4e^{-|4t|}dt = 2\int_{0}^{T} 4e^{-4t}dt$$

$$= \frac{8e^{-4t}}{-4}\bigg|_{0}^{T} = 2(1 - e^{-4T})$$

$$\lim_{T \to \infty} 2(1 - e^{-4T}) = 2$$

Since the limit is finite this is an energy signal

Power Signal

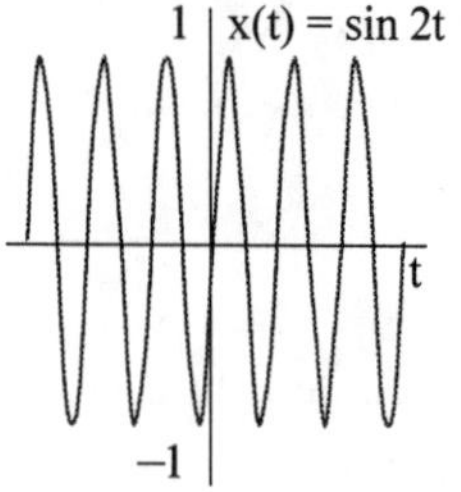

$$\frac{1}{2T}\int_{-T}^{T} \sin^2 2t\,dt = \frac{1}{T}\int_{0}^{T} \sin^2 2t\,dt$$

$$= \frac{1}{2}\frac{1}{T}\int_{0}^{T}(1 - \cos 4t)dt$$

$$= \frac{1}{2}\frac{1}{T}\left(T - \frac{\sin 4T}{4}\right)$$

and $\lim_{T \to \infty}\left(\frac{1}{2} - \frac{\sin 4T}{8T}\right) = \frac{1}{2}$

and this is a power signal.

Correlation Functions
Energy Signals x(t), y(t), x(n), y(n)

Autocorrelation functions for x(t), x(n)

$$R_x(\tau) = \int_{-\infty}^{\infty} x(t)\, x(t+\tau)\, dt \quad : \text{continuous time}$$

$$R_x(h) = \sum_{n=-\infty}^{\infty} x(n)\, x(n+h) : \text{discrete time} \tag{5}$$

Cross correlation functions between x(t) and y(t), x(n) and y(n)

$$R_{xy}(\tau) = \int_{-\infty}^{\infty} x(t)\, y(t+\tau)\, dt \quad : \text{continuous time}$$

$$R_{xy}(h) = \sum_{n=-\infty}^{\infty} x(n)\, y(n+h) : \text{discrete time} \tag{6}$$

Power Signals x(t), y(t), x(n), y(n)

Autocorrelation functions for x(t), x(n)

$$R_x(t) = \lim_{T\to\infty} \int_{-T}^{T} x(t)\, x(t+\tau)\, dt \quad : \quad \text{continuous}$$

$$R_x(h) = \lim_{N\to\infty} \frac{1}{2N+1} \sum_{n=-N}^{N} x(n)\, x(n+h) : \text{discrete} \tag{7}$$

Cross correlation functions between x(t) and y(t), x(n) and y(n)

$$R_{xy}(t) = \lim_{T\to\infty} \int_{-T}^{T} x(t)\, y(t+\tau)\, dt : \quad \text{continuous}$$

$$R_{xy}(h) = \lim_{N\to\infty} \frac{1}{2N+1} \sum_{n=-N}^{N} x(n)\, y(n+h) : \text{discrete} \tag{8}$$

4. CONTINUOUS TIME CONVOLUTION

1. Definitions
   (a) h(t) and x(t) defined for all t

$$y(t) = \int_{-\infty}^{\infty} x(\tau)h(t-\tau)d\tau = \int_{-\infty}^{\infty} x(t-\tau)h(\tau)d\tau$$

   (b) causal h(t): { h(t) = 0, t < 0} {Note limits}

$$y(t) = \int_{-\infty}^{t} x(\tau)h(t-\tau)d\tau = \int_{0}^{\infty} x(t-\tau)h(\tau)d\tau$$

   (c) causal h(t), x(t): h(t), x(t) = 0, t < 0 {Note limits}

$$y(t) = \int_{0}^{t} x(\tau)h(t-\tau)d\tau = \int_{0}^{t} x(t-\tau)h(\tau)d\tau$$

2. Step convolution

$$y(t) = \int_{-\infty}^{\infty} x(\tau)u(t-\tau-t_0)d\tau = \int_{-\infty}^{t-t_0} x(\tau)d\tau$$

3. Impulse convolution

$$y(t) = \int_{-\infty}^{\infty} x(\tau)\,\delta(t-\tau-t_0)d\tau$$

$$= \int_{-\infty}^{\infty} x(t-\tau)\delta(\tau-t_0)d\tau = x(t-t_0)$$

4. Doublet convolution : $\dot{\delta}(t) = \dfrac{d\delta(t)}{dt}$

$$y(t) = \int_{-\infty}^{\infty} x(\tau)\frac{d}{dt}[\delta(t-\tau)]d\tau$$

$$= \frac{d}{dt}\int_{-\infty}^{\infty} x(\tau)[\delta(t-\tau)]d\tau = \frac{d}{dt}x(t)$$

5. Commutative law

$$y(t) = \int_{-\infty}^{\infty} x(\tau)h(t-\tau)d\tau = \int_{-\infty}^{\infty} x(t-\tau)h(\tau)d\tau$$

6. Distributive law

$$y(t) = x(t) * [h_1(t)+h_2(t)] = x(t) * h_1(t) + x(t) * h_2(t)$$

7. Associative law

$$[x(t) * h_1(t)] * h_2(t) = x(t) * [h_1(t) * h_2(t)]$$

8. Time Invariance

If $y(t) = x(t) * h(t)$ then $y(t-t_0) = x(t-t_0) * h(t)$

$$= x(t) * h(t-t_0)$$

9. Campbell's theorem

(a) Pulse-width of convolution product is equal to the sum of the pulse-widths of convolved functions.
(b) Delay of convolution product is equal to the sum of the delays of convolved functions.
(c) Area of the convolution product is equal to the product of the areas of the convolved functions.

10. Scaling

If $y(t) = x(t) * h(t)$ then $|a|\, y\left(\frac{t}{a}\right) = x\left(\frac{t}{a}\right) * h\left(\frac{t}{a}\right)$

5. DISCRETE LINEAR AND CIRCULAR CONVOLUTION

Linear Discrete Convolution

1. Definitions

(a) h(n) and x(n) defined for all n

$$y(n) = \sum_{m=-\infty}^{\infty} x(m)h(n-m) = \sum_{m=-\infty}^{\infty} x(n-m)h(m)$$

(b) causal h(n): { h(n) = 0, n < 0} {Note limits}

$$y(n) = \sum_{m=-\infty}^{n} x(m)h(n-m) = \sum_{m=0}^{\infty} x(n-m)h(m)$$

(c) causal h(n), x(n): { h(n), x(n) = 0, n < 0}

$$y(n) = \sum_{m=0}^{n} x(m)h(n-m) = \sum_{m=0}^{n} x(n-m)h(m)$$

2. Discrete step convolution

$$y(n) == \sum_{m=-\infty}^{\infty} x(m)u(n-m-n_0) = \sum_{n=-\infty}^{n-n_0} x(n)$$

3. Discrete impulse (Kronecker delta) convolution

$$y(n) == \sum_{m=-\infty}^{\infty} x(m)\delta(n-m-n_0) = x(n-n_0)$$

4. Discrete convolution also follows the commutative, distributive, associative and scaling laws as in the continuous case.

5. Time Invariance

If $y(n) = x(n) * h(n)$ then $y(n - n_0) = x(n - n_0)*h(n)$
$$= x(n)*h(n - n_0)$$

6. Campbell's theorem (applies only to linear discrete convolution)

(a) The number of points in the convolution product is equal to one less than the sum of the number of points in the convolved sequences.

(b) Delay in the convolution product is equal to the sum of the delays of the convolved sequences.

(c) Sum of points in the convolution product is equal to the product of the sums of points in the convolved sequences.

Circular Discrete Convolution

1. Definition: x(n) and h(n) are periodic sequences with the *same period* N

$$y(n) = \sum_{m=0}^{N-1} x(m)h(n - m) = \sum_{m=0}^{N-1} x(n - m)h(m)$$
$$= x(n) \circledast h(n) = y(n + kN)\ , k = \ldots, -1, 0, 1, \ldots$$

2. If the periods of x(n) and h(n) are not the same, zeros can be padded to either of them so that they have the same period and circular convolution can be performed.

3. Circular discrete convolution also follows the commutative, distributive, associative, and scaling laws.

4. The period of the convolution product is the same as the period of the convolved sequences.

5. Sum of points in a period of the convolution product is equal to the product of the sums of points in a period of the convolved sequences.

6. Circular convolution of two sequences can be made to correspond to the linear convolution in one period, if the period of the convolved sequences is made up to have the same number of points by zero padding as the number in the linear convolution.

Example - Linear Discrete Convolution

Input sequence

$x(n) = \{1, 1, 1, 1, 1\}$

Transfer sequence

$h(n) = \{0, 0, 1, 1, 1, 1, 1, 1, 0, 0, 0, 1, 1, 1, 1, 1, 1\}$

Output sequence

$$y(n) = x(n)*h(n) = \sum_{n=0}^{20} x(m)\, h(n-m)$$

The convolution diagram is shown on the next page.

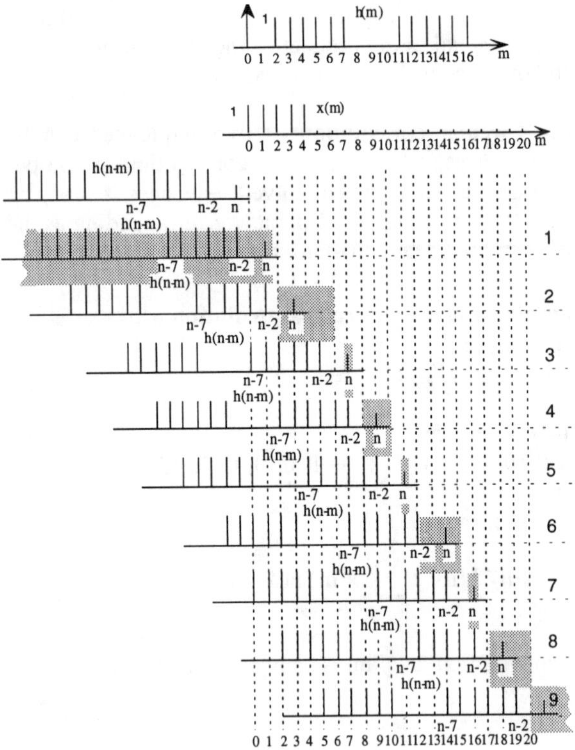

From the figure, the ranges of n, m and the values of y(n) are shown in the table below:

$n < 2$	$m = 0$	$y(n) = 0$
$2 \le n \le 6$	$0 \le m \le n-2$	$1 \le y(n) \le 5$
$n = 7$	$0 \le m \le 4$	$y(n) = 5$
$8 \le n \le 10$	$n-7 \le m \le 4$	$4 \ge y(n) \ge 2$
$n = 11$	$m = 0$ and 4	$y(n) = 2$

$12 \le n \le 15$	$0 \le m \le n-11$	$2 \le y(n) \le 5$
$n = 16$	$0 \le m \le 4$	$y(n) = 5$
$17 \le n \le 20$	$n-16 \le m \le 4$	$4 \ge y(n) \ge 1$
$n > 20$	$m = 0$	$y(n) = 0$

The corresponding plot of y(n) is also shown in diagram

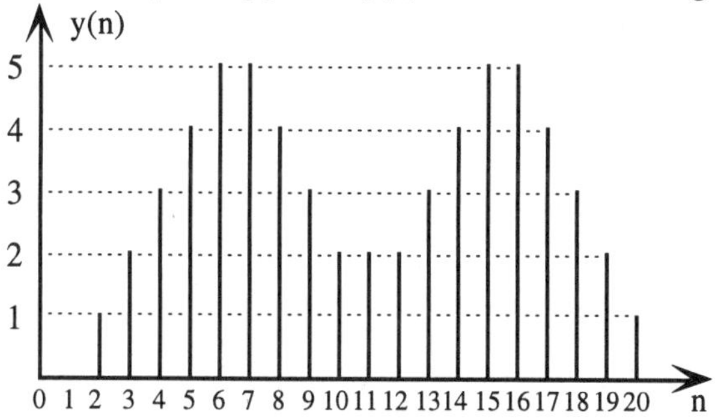

EXAMPLE - Circular Convolution Machine

Input sequence

$x(n) = \{1, 1, 1, 1, 1\}$

Transfer sequence

$h(n) = \{0, 0, 1, 1, 1, 1, 1, 1, 0, 0, 0, 1, 1, 1, 1, 1, 1\}$

Padding x(n) with 12 zeros to make it equal to the number of points in h(n), we obtain,

$x(n) = \{1, 1, 1, 1, 1, 0, 0, 0, 0, 0, 0, 0, 0, 0, 0, 0, 0\}$

$h(n) = \{0, 0, 1, 1, 1, 1, 1, 1, 0, 0, 0, 1, 1, 1, 1, 1, 1\}$

The circular convolution of x(n) with h(n) is:

$$y(n) = x(n) \circledast h(n) = \sum_{n=0}^{16} x(m)h(n-m)$$

$$= y(n + 17k),\ k = \ldots, -1, 0, 1\ldots$$

The corresponding diagram is shown on the next page.

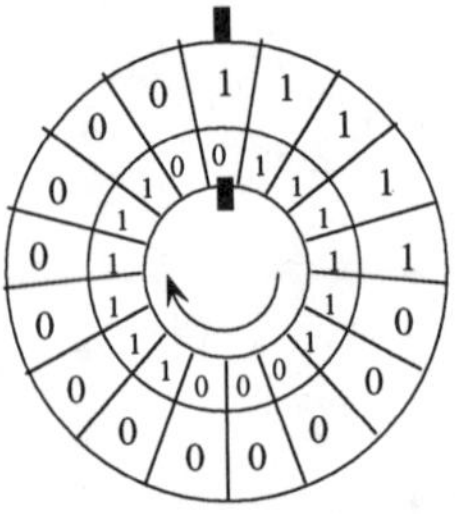

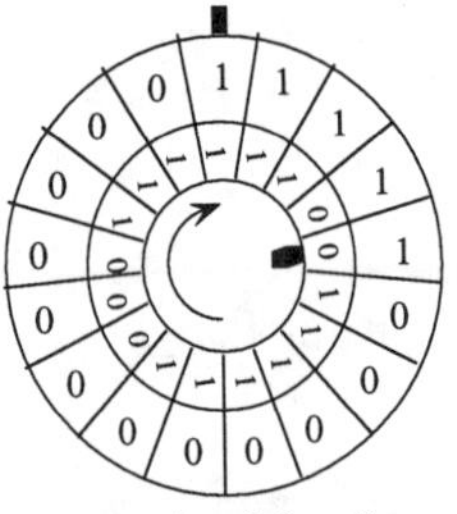

n = 0, y(n) = 4

n = 4, y(n) = 3

Note the time markers on the inner and outer wheels in all the four positions

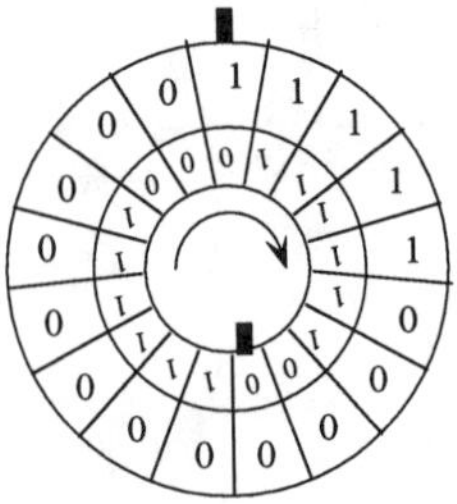

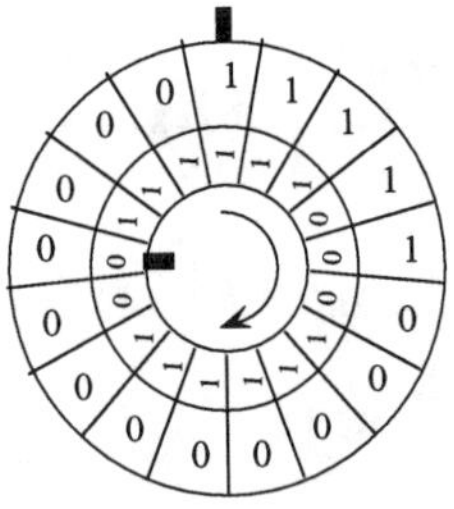

n = 8, y(n) = 4

n = 13, y(n) = 3

Using the circular convolution machine, the values of y(n) for every value of n can be plotted. Since the inner circle can go round the outer circle indefinitely, y(n) is periodic with period N = 17. y(n) is shown in the diagram on the next page for values of n in the fundamental period $0 \le n \le 16$.

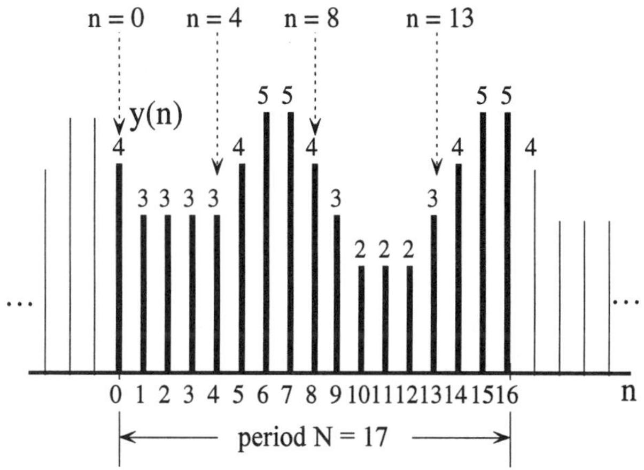

Example - Circular Discrete Convolution to Linear Discrete Convolution

Input sequence

x(n) = {1, 1, 1, 1, 1}

Transfer sequence

h(n) = {0, 0, 1, 1, 1, 1, 1, 1, 0, 0, 0, 1, 1, 1, 1, 1, 1}

Pad x(n) and h(n) with zeros such that the # of points in each of them is equal to the # of points in linear convolution as shown below.

x(n) = {1, 1, 1, 1, 1, 0, 0, 0, 0, 0, 0, 0, 0, 0, 0, 0, 0, 0, 0, 0, 0}

h(n) = {0, 0, 1, 1, 1, 1, 1, 1, 0, 0, 0, 1, 1, 1, 1, 1, 1, 0, 0, 0, 0}

In this case the total number of points = 21. y(n) is obtained from the circular convolution machine with period 21 and the linear convolution part is the fundamental period, $0 \le n \le 21$, as shown in the diagram. The circular convolution is given by:

$$y(n) = x(n) \circledast h(n) = \sum_{\nu=0}^{20} x(m)h(n-m)$$

$$= y(n + 21k),\ k = \ldots, -1, 0, 1\ldots$$

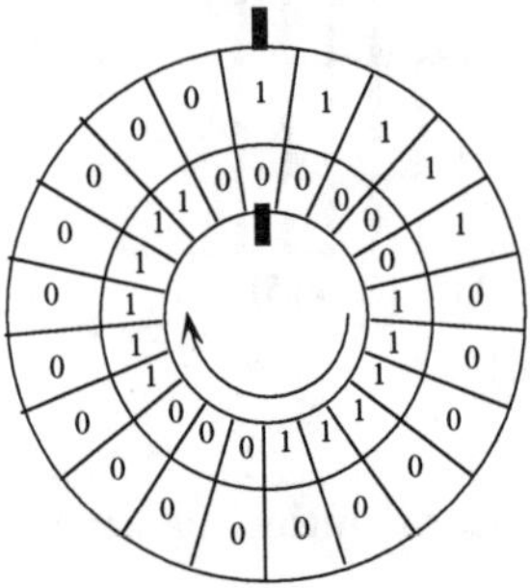

n = 0, y(n) = 0

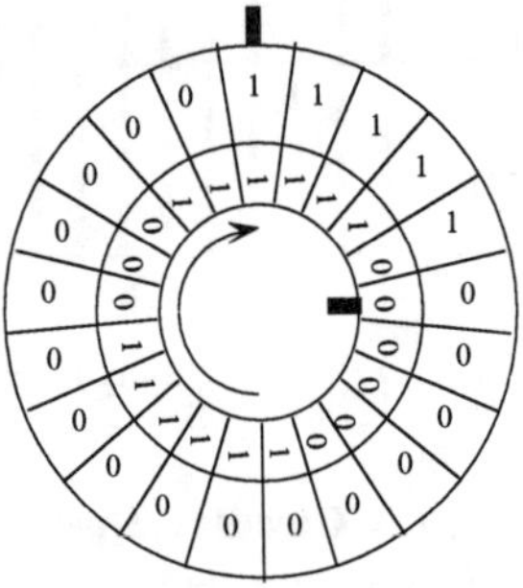

n = 5, y(n) = 4

Note the time markers on the inner and outer wheels in all the four positions

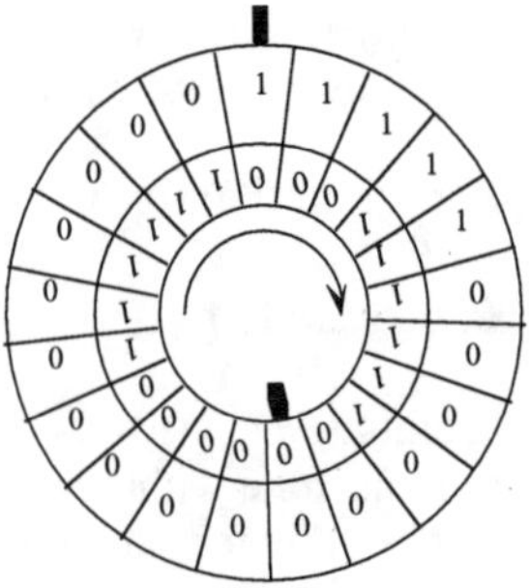

n = 10, y(n) = 2

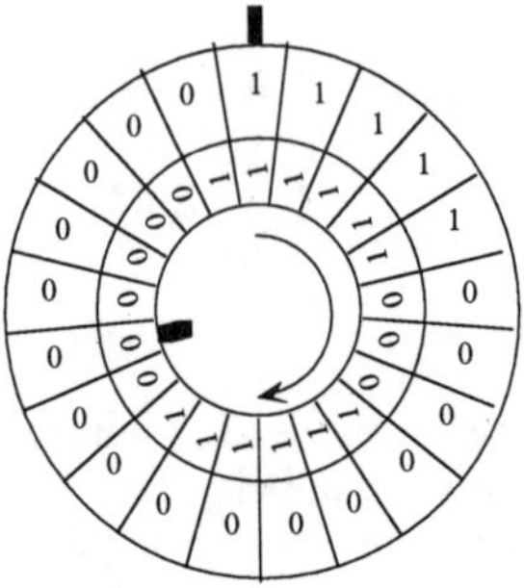

n = 15, y(n) = 5

The plot of y(n) is shown in the diagram for the fundamental period $0 \le n \le 21$. Clearly, y(n) is periodic with period N = 21.

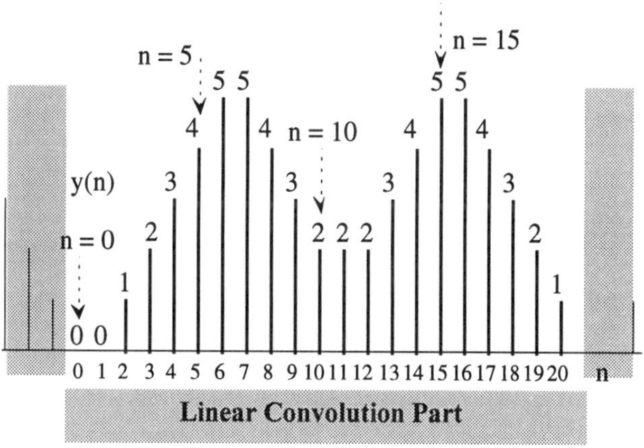

6. EIGENFUNCTIONS AND ORTHOGONAL POLYNOMIALS

A general second order linear differential operator D is of the general form

$$D\,f(x) = \left[A(x)\frac{d^2}{dx^2} + C(x)\frac{d}{dx} + B(x) \right] f(x) \qquad (1)$$

The real functions A(x), B(x), C(x) are defined over the region of interest [a, b] and are continuously differentiable. The zeros of A(x) are the singular points of the differential operator D. We choose the region [a, b] such that A(x) is not equal to zero in the interior of the region [a, b]. We now define an *adjoint* operator $\overline{D}$ by

$$\overline{D}\,f(x) = \frac{d^2}{dx^2}\Big(A(x)\,f(x)\Big) - \frac{d}{dx}\Big(C(x)f(x)\Big) + B(x)\,f(x)$$

$$= A(x)\frac{d^2\,f(x)}{dx^2} + \Big(2\,A(x) - C(x)\Big)\frac{d\,f(x)}{dx} \qquad (2)$$

$$+ \left(\frac{d^2\,A(x)}{dx^2} - \frac{d\,q(x)}{dx} + B(x) \right) f(x)$$

A necessary and sufficient condition for the linear differential operators D and $\overline{D}$ to be equal is,

$$\frac{d\,A(x)}{dx} = C(x) \qquad (3)$$

When this condition is satisfied then we have,

$$\overline{D}\,f(x) = D\,f(x) = \frac{d}{dx}\left[A(x)\frac{d\,f(x)}{dx} \right] + B(x)\,f(x) \qquad (4)$$

and the differential operator D is said to be *self-adjoint* and D is called the Sturm-Liouville differential operator.

Eigenfunctions (orthogonal polynomials) in the interval [a, b] like Legendre, LaGuerre, Hermite, Bessel, etc., are obtained as solutions to second order linear differential equations. Most of these eigenfunctions can be characterized by the Sturm-Liouville differential operator, D, given in eq.(4). Starting from the Sturm-Liouville differential operator we can derive the properties of eigenvalues and eigenfunctions and how these eigenfunctions can form the basis vectors of an orthogonal function space. Let us take two functions, u(x) and v(x), defined in the interval $x \in [a, b)$. It can be readily shown that a total differential can be formed from $[v(x)Du(x) - u(x)Dv(x)]$ as shown below:

$$v(x)Du(x) - u(x)Dv(x) = \frac{d}{dx}\left[A(x)\left(v(x)\frac{du(x)}{dx} - u(x)\frac{dv(x)}{dx}\right)\right] \tag{5}$$

If eq.(5) is integrated in the interval [a, b), there results

$$\int_a^b \left[v(x)Du(x) - u(x)Dv(x)\right] dx = \left[A(x)\left(v(x)\frac{du(x)}{dx} - u(x)\frac{dv(x)}{dx}\right)\right]\Bigg|_a^b \tag{6}$$

If the boundary conditions are such that the right-hand side of eq.(6) is zero, i.e.,

$$\left[A(x)\left(v(x)\frac{du(x)}{dx} - u(x)\frac{dv(x)}{dx}\right)\right]_a^b = 0 \tag{7}$$

then the left-hand side of eq.(6) is

$$\int_a^b \left(v(x)Du(x) - u(x)Dv(x)\right) dx = 0 \tag{8}$$

For a Sturm-Liouville differential equation given by $D\phi(x) = \lambda\, \psi(x)$

$$\text{Or, } \frac{d}{dx}\left(A(x)\frac{d}{dx}\right)\psi(x) + \left(B(x) - \lambda\right)\psi(x) = 0 \tag{9}$$

only a selected number of $\{\lambda_i\}$ satisfies eq.(9) with the imposed boundary conditions governed by eq.(7). These selected $\{\lambda_i\}$ are called *eigenvalues* and the corresponding solutions $\{\psi_\iota(x)\}$ are called the *unnormalized eigenfunctions*. The eigenvalues for eq.(9) are all real distinct, and the eigenfunctions form a linearly independent set. However, by virtue of eqs.(7), (8) the eigenfunctions $\{\psi_i(x)\}$ are not only linearly independent but also orthogonal as we will show presently. Substituting $u(x) = \psi_i(x)$ and $v(x) = \psi_k(x)$ in eq.(8) we obtain,

$$\int_a^b \big(\psi_k(x)D\psi_i(x) - \psi_i(x)D\psi_k(x)\big)\, dx = 0 \tag{10}$$

Substituting $D\psi_i(x) = \lambda_i\, \psi_i(x)$ and $D\psi_k(x) = \lambda_k\, \psi_k(x)$ yields

$$\int_a^b \big(\psi_k(x)\lambda_i\, \psi_i(x) - \psi_i(x)\lambda_k\, \psi_k(x)\big)\, dx = 0,\ i \neq k$$

Or, $$\int_a^b \big[\psi_i(x)\psi_k(x)(\lambda_i - \lambda_k)\big]\, dx = 0,\ i \neq k \tag{11}$$

Since $\lambda_i \neq \lambda_k$ eq.(11) gives the orthogonality condition

$$\int_a^b \big[\psi_i(x)\psi_k(x)\big]\, dx = 0,\ i \neq k \tag{12}$$

The unnormalized eigenfunctions $\{\psi_i(x)\}$ can be normalized by setting

$$\int_a^b \big(\psi_i(x)\big)^2\, dx = \lambda_i^2 \quad i = 1, 2, \ldots, k, \ldots \tag{13}$$

and defining the normalized eigenfunctions as

$$\phi_i(x) = \frac{\psi_i(x)}{\lambda_i} \quad i = 1, 2, \ldots, k, \ldots \quad \text{so that}$$

$$\int_a^b \big(\phi_i(x)\big)^2\, dx = 1 \quad i = 1, 2, \ldots, k, \ldots \tag{14}$$

and the sequence $\{f_i(x)\}$ can be used as the basis vectors of an orthogonal function space.

Depending on the form of the differential equation (1), we can obtain a variety of orthonormal functions. Any arbitrary second order linear differential equation of the form,

$$p(x)\frac{d^2 f(x)}{dx^2} + q(x)\frac{df(x)}{dx} + r(x)\, f(x) = \lambda\, f(x) \qquad (15)$$

can be expressed in the Sturm-Liouville form given by eq.(9). We first multiply both eq.(15) by a suitable weighting function w(x) to yield

$$w(x)\, p(x)\frac{d^2 f(x)}{dx^2} + w(x)\, q(x)\frac{df(x)}{dx} + w(x)\, r(x)\, f(x) = w(x)\, \lambda\, f(x) \qquad (16)$$

To convert eq.(16) into the desired form we set from eq.(3)

$$\frac{d}{dx}\Big[w(x)\, p(x)\Big] = w(x)\, q(x) \quad \text{Or,}$$

$$\frac{1}{w(x)}\frac{dw(x)}{dx} = \frac{q(x)}{p(x)} - \frac{1}{p(x)}\frac{dp(x)}{dx} \qquad (17)$$

If we now identify A(x) in the operator D of eq.(1) as A(x) = w(x)p(x), and B(x) = w(x)r(x), then eq.(16) can be expressed in the desired Sturm-Liouville form eq.(9) as,

$$\frac{d}{dx}\left[A(x)\frac{d\psi(x)}{dx}\right] + B(x)\, \psi(x) = \lambda\, w(x)\, \psi(x) \qquad (18)$$

The orthogonality condition corresponding to eq.(12) now becomes,

$$\int_a^b w(x)\,\psi_i(x)\,\psi_k(x)\,dx = 0 \qquad i \neq k \tag{19}$$

Eq.(19) is known as the weighted orthogonality condition.

In the same manner as eq.(13) we can normalize $\{\psi(x)\}$ by noting that,

$$\int_a^b w(x)\,\psi_i^2(x)\,dx = \lambda_i^2 \qquad i = 1, 2, \ldots, k, \ldots \tag{20}$$

and setting

$$\phi_i(x) = \frac{\psi_i(x)}{\lambda_i} \quad i = 1, 2, \ldots, k, \ldots \text{ so that}$$

$$\int_a^b w(x)\left(\phi_i(x)\right)^2 dx = 1 \quad i = 1, 2, \ldots, k, \ldots \tag{21}$$

The weighting function w(x) can be determined by integrating both sides of eq.(17). Or,

$$\int \frac{1}{w(x)}\frac{dw(x)}{dx}\,dx = \int \frac{q(x)}{p(x)} - \frac{1}{p(x)}\frac{dp(x)}{dx}\,dx$$

$$\ln w(x) = \int \frac{q(x)}{p(x)}\,dx - \ln p(x)$$

$$w(x) = e^{\ln\left(1/p(x)\right)}\, e^{\int \frac{q(x)}{p(x)}\,dx}$$

Or, $$w(x) = \frac{e^{\int \frac{q(x)}{p(x)}\,dx}}{p(x)} \tag{22}$$

We shall now apply the above techniques to a hypergeometric differential equation given by,

$$x(1-x)\frac{d^2 f(x)}{dx^2} + \left[\gamma - \left(1+\alpha+\beta\right)x\right]\frac{d f(x)}{dx} - \alpha\,\beta\, f(x) = 0 \qquad (23)$$

The differential equation (23) has regular interior singular points given by the solution to $x(1-x) = 0$. Using the Sturm-Liouville techniques as discussed above, we shall find series solutions in the neighborhood of $x = 0$. We shall first determine the weighting function $w(x)$ that will transform eq.(23) into the Sturm-Liouville form. From eq.(22)

$$\begin{aligned}\int \frac{q(x)}{p(x)}\,dx &= \int \frac{\gamma}{x(1-x)}\,dx - \int \frac{1+\alpha+\beta}{1-x}\,dx \\ &= \int \frac{\gamma}{x}\,dx + \int \frac{\gamma}{1+x}\,dx - \int \frac{1+\alpha+\beta}{1-x}\,dx \\ &= \gamma \ln(x) - \gamma \ln(1-x) + (1+\alpha+\beta)\ln(1-x) \\ &= \gamma \ln(x) + (1+\alpha+\beta-\gamma)\ln(1-x)\end{aligned}$$

and

$$w(x) = \frac{e^{\int (q(x)/p(x))\,dx}}{p(x)} = \frac{e^{\gamma \ln(x) + (1+\alpha+\beta-\gamma)\ln(1-x)}}{x(1-x)} = x^{\gamma-1}\,(1-x)^{\alpha+\beta-\gamma} \qquad (24)$$

Solution about regular singular points

We shall first find a solution $f(x)$ about the regular singular point $x = 0$. The procedure is to seek solutions of the form,

$$f(x) = \sum_{k=0}^{\infty} a_k\, x^{r+k} \qquad (25)$$

In eq.(25) we have to determine

a. The values for r for which eq.(23) has a solution.
b. The recurrence relationship for a_k.
c. The region of convergence.

We substitute eq.(25) into eq.(23) noting that,

$$\frac{df(x)}{dx} = \sum_{k=0}^{\infty} a_k \, (r+k)\, x^{r+k-1}$$

$$\frac{d^2 f(x)}{dx^2} = \sum_{k=0}^{\infty} a_k \, (r+k)(r+k-1)\, x^{r+k-2} \tag{26}$$

The resulting equation is,

$$\sum_{k=0}^{\infty} \Big\{ \big[(r+k)(r+k-1) + \gamma\,(r+k)\big]\, a_k\, x^{r+k-1} - \big[(r+k)(r+k-1) + (r+k)\,(1+\alpha+\beta) + \alpha\,\beta\big]\, a_k\, x^{r+k} \Big\} = 0 \tag{27}$$

The first term in eq.(27) can be written in the form

$$\sum_{k=0}^{\infty} \big[(r+k)(r+k-1) + \gamma\,(r+k)\big]\, a_k\, x^{r+k-1}$$

$$= \sum_{k=-1}^{\infty} \big[(r+k+1)(r+k) + \gamma\,(r+k+1)\big]\, a_{k+1}\, x^{r+k} \tag{28}$$

Eq.(28) is now substituted into eq.(27) to yield

$$r\,(r-1+\gamma)\, a_0\, x^{r-1} + \sum_{k=0}^{\infty} \Big\{ \big[(r+k+1)(r+k+\gamma) a_{k+1}\big]$$

$$-\left[\left((r+k)(r+k+\alpha+\beta)+\alpha\,\beta\right)a_k\right]\}\,x^{r+k}=0 \tag{29}$$

If eq.(29) is to be true, then each term must be equal to zero. Thus,

$$r\left((r-1+\gamma\right)=0 \tag{30}$$

$$\begin{aligned}(r+k+1)(r+k+\gamma)\,a_{k+1}\\ =\left[(r+k)(r+k+\alpha+\beta)+\alpha\,\beta\right]a_k\end{aligned} \tag{31}$$

Eq.(30) is called the indicial equation (similar to the characteristic equation in linear constant coefficient systems) that gives rise to the values of r for which solutions may exist. We now have two independent solutions corresponding to the roots $r = 0$ and $r = 1 - \gamma$, and these are

$$f_1(x)=\sum_{k=0}^{\infty} a_k\,x^k \tag{32}$$

$$f_2(x)=x^{1-\gamma}\sum_{k=0}^{\infty} a_k\,x^k \tag{33}$$

provided $(1 - \gamma)$ is neither zero nor an integer. If $(1 - \gamma)$ is an integer, then eq.(33) is not independent of eq.(32); in which case we have to determine the second solution by other means.

Eq.(31) gives the recursive relation for the coefficients $\{a_k\}$as given by,

$$a_{k+1}=\left[\frac{(r+k)(r+k+\alpha+\beta)+\alpha\,\beta}{(r+k+1)(r+k+\gamma)}\right]a_k,\ k>0 \tag{34}$$

yielding two different sets of recursive relationships corresponding to $r = 0$ and $r = (1 - \gamma)$. We can determine the region of convergence(ROC) for the series in eq.(25). The easiest way is to apply the ratio test and determine the condition under which the ratio is less than 1 as $n \to \infty$. Applying the ratio test for f(x) we obtain,

$$\lim_{k \to \infty} \left| \frac{a_{k+1} x^{k+r+1}}{a_k x^{k+r}} \right|$$

$$= \lim_{k \to \infty} \left| \frac{(r+k)(r+k+\alpha+\beta) + \alpha\beta}{(r+k+1)(r+k+\gamma)} x \right| = |x| < 1$$

The series converges for $-1 < x < 1$ for all values of r.

We shall now find the recursive relation for the coefficients $\{a_k\}$ corresponding to $r = 0$. Under this condition, eq.(34) becomes,

$$a_{k+1} = \frac{(k+\alpha)(k+\beta)}{(k+1)(k+\gamma)} a_k, \ k > 0 \qquad (35)$$

Hence, assuming that the constant $a_0 = 1$ and $\gamma > 0$ the first solution $f_1(x)$ is given by,

$$f_1(x)$$

$$= 1 + \frac{\alpha\beta}{1.\gamma} x + \frac{\alpha(\alpha+1)\beta(\beta+1)}{2!\ \gamma.(\gamma+1)} x^2 +$$

$$\cdots$$

$$\cdots$$

$$+ \frac{\alpha(\alpha+1) \ldots (\alpha+k-1)\ \beta(\beta+1) \ldots (\beta+k-1)}{k!\ \gamma.(\gamma+1) \ldots (\gamma+k-1)} x^k + \ldots \qquad (36)$$

In a similar manner we can determine the second solution, $f_2(x)$, under the condition that $1 - \gamma$ is neither an integer nor equal to zero as,

$$f_2(x)$$

$$= x^{(1-\gamma)} \left\{ 1 + \frac{(\alpha-\gamma+1)(\beta-\gamma+1)}{1.(2-\gamma)} x \right.$$

$$+\frac{(\alpha-\gamma+1)(\alpha-\gamma+2)(\beta-\gamma+1)(\beta-\gamma+2)}{2!\,(2-\gamma)\,(3-\gamma)}\,x^2$$

$$\cdots$$

$$\cdots$$

$$+\left[\frac{(\alpha-\gamma+1)(\alpha-\gamma+2)\,\ldots\,(\alpha-\gamma+k)}{k!\,(2-\gamma)(3-\gamma)\,\ldots\,(k+1-\gamma)}\times \frac{(\beta-\gamma+1)(\beta-\gamma+2)\,\ldots\,(\beta-\gamma+k)}{}\right]x^k+\ldots\Bigg\} \tag{37}$$

Eqs.(36) and (37) are infinite series but they terminate into a polynomial for suitable values of α and β. In eq.(35) if we substitute $\alpha = -n$, then a_{n+1} is zero and all the higher coefficients a_{n+k} are also zero. Thus, the solution to the differential eq.(23) is a polynomial. For computational ease the coefficient of $\frac{df(x)}{dx}$ is usually independent of n, and in that case we substitute $\beta = n + \gamma + \delta - 1$, where we have defined another constant $\delta > 0$. With these values of α and β, the differential eq.(23) now becomes,

$$x(1-x)\,\frac{d^2 f(x)}{dx^2} + \left[\gamma - (\gamma+\delta)\,x\right]\frac{d\,f(x)}{dx} + n\,(n+\gamma+\delta-1)\,f(x) = 0 \tag{38}$$

The recursive eq.(35) now becomes

$$a_{k+1} = \left[\frac{(r+k)(r+k+\gamma+\delta-1) - n\,(n+\gamma+\delta-1)}{(r+k+1)(r+k+\gamma)}\right] a_k,\ k > 0$$

Under the condition $r = 0$, the recursive solution for a_{k+1} becomes,

$$a_{k+1} = \left[\frac{(k-n)(k+n+\gamma+\delta-1)}{(k+1)(k+\gamma)}\right] a_k, \; k > 0 \qquad (39)$$

with the coefficients greater than a_n being equal to zero. We can now write the coefficients $\{a_k\}$ from eq.(39),

$$a_0 = 1$$

$$a_1 = (-1)^1 \frac{n(n+\gamma+\delta-1)}{\gamma}$$

$$a_2 = (-1)^2 \frac{n(n-1)(n+\gamma+\delta-1)(n+\gamma+\delta)}{\gamma(\gamma+1)} \frac{1}{2!}$$

$$a_3 = (-1)^3 \left[\frac{n(n-1)(n-2)(n+\gamma+\delta-1)(n+\gamma+\delta)}{\gamma(\gamma+1)(\gamma+2)} \times (n+\gamma+\delta+1)\right] \frac{1}{3!}$$

$$\vdots$$

$$a_k = (-1)^k \left[\frac{n(n-1)(n-2)\ldots(n-k+1)}{\gamma(\gamma+1)\ldots(\gamma+k-1)} \times (n+\gamma+\delta-1)(n+\gamma+\delta)\ldots(n+\gamma+\delta+k-2)\right] \frac{1}{k!}$$

$$\vdots$$

$$a_n = (-1)^n \frac{n!(n+\gamma+\delta-1)\ldots(2n+\gamma+\delta-2)}{\gamma(\gamma+1)\ldots(\gamma+n-1)} \frac{1}{n!} \qquad (40)$$

and the corresponding series solution for $f(x)$ is called the Jacobi polynomials $P_n^{(\gamma,\delta)}(x)$, with $P_0(x) = 1$. These polynomials are given by,

$$P_n^{(\gamma,\delta)}(x) = 1 + \sum_{k=1}^{n} \left[\frac{n(n-1)(n-2)\ldots(n-k+1)}{\gamma(\gamma+1)\ldots(\gamma+k-1)}\right] \times$$

$$\times \left[(n+\gamma+\delta-1)(n+\gamma+\delta)\ldots(n+\gamma+\delta+k-2)\right] \frac{(-x)^k}{k!} \quad (41)$$

for n taking integral values 1, 2 …, and they are orthogonal in the interval [0, 1].

Example

As an example, if $\gamma = 3/2$ and $\delta = 2$, then eq.(41) can simplified to,

$$P_n^{(3/2,2)}(x) = 1 + \sum_{k=1}^{n} \frac{[n(n-1)(n-2)\ldots(n-k+1)]}{\frac{3}{2}\frac{5}{2}\ldots\left(\frac{1}{2}+k\right)} \times \quad (42)$$

$$\times \left[(n+\frac{5}{2})(n+\frac{7}{2})\ldots(n+\frac{3}{2}+k)\right] \frac{(-x)^k}{k!}$$

The first five Jacobi polynomials are shown below. They are not orthonormal and have to be normalized.

$$P_0^{(3/2,2)}(x) = 1$$

$$P_1^{(3/2,2)}(x) = 1 - \frac{7}{3}x$$

$$P_2^{(3/2,2)}(x) = 1 - 6x + \frac{33}{5}x^2$$

$$P_3^{(3/2,2)}(x) = 1 - 11x + \frac{143}{5}x^2 - \frac{143}{7}x^3$$

$$P_4^{(3/2,\,2)}(x) = 1 - \frac{52}{3}\,x + 78\,x^2 - \frac{884}{7}\,x^3 + \frac{4199}{63}\,x^4 \qquad (43)$$

The plots of these five polynomials are shown below.

JACOBI POLYNOMIALS

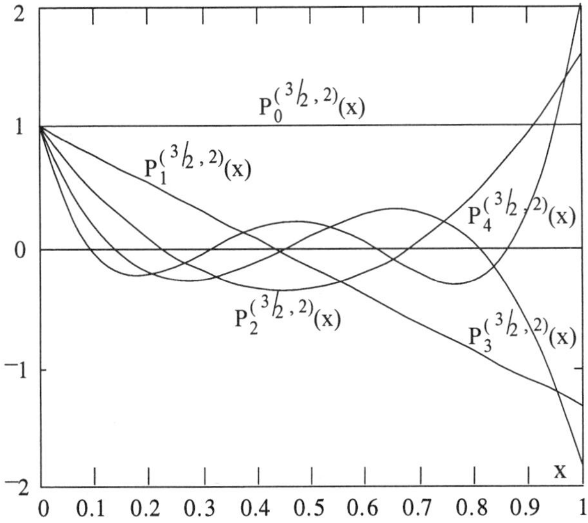

Solution about an ordinary point

In the last example we formed a series solution around a singular point of the differential equation to obtain orthogonal polynomials. We shall now form series solutions about an ordinary point using the Legendre differential equation given by,

$$\left(1 - x^2\right)\frac{d^2 f(x)}{dx^2} - 2\,x\,\frac{df(x)}{dx} + n(n+1)\,f(x) = 0 \qquad (44)$$

The singular points are at $x = \pm 1$. It can be shown that the weighting function $w(x) = 1$. We shall now obtain series solutions about the ordinary point $x = 0$. In this case we seek solutions of the form,

$$f(x) = \sum_{k=0}^{\infty} a_k x^k \quad \text{with} \tag{45}$$

$$\frac{df(x)}{dx} = \sum_{k=1}^{\infty} a_k k x^{k-1}$$

$$\frac{d^2 f(x)}{dx^2} = \sum_{k=2}^{\infty} a_k k (k-1) x^{k-2} \tag{46}$$

Substituting eqs.(45) and (46) into eq.(44) and changing the index of summation we obtain,

$$\sum_{k=0}^{\infty} a_{k+2} (k+2)(k+1) x^k - \sum_{k=2}^{\infty} a_k k (k-1) x^k$$
$$-2 \sum_{k=1}^{\infty} a_k k x^k + n(n+1) \sum_{k=0}^{\infty} a_k x^k = 0$$

Or,

$$2 a_2 + n(n+1) a_0 + \left[6 a_3 - \left(2 - n(n+1)\right) a_1\right] x$$
$$+ \sum_{k=2}^{\infty} \left\{ (k+2)(k+1) a_{k+2} \right. \tag{47}$$
$$\left. - \left[k(k-1) + 2k - n(n+1)\right] a_k \right\} x^k = 0$$

Setting each power of x in eq.(47) to zero, there results,

$$a_2 = -\frac{n(n+1)}{2} a_0 : \; a_3 = \frac{\left(2 - n(n+1)\right)}{6} a_1 \tag{48}$$

$$a_{k+2} = \frac{[k(k+1) - n(n+1)]}{(k+2)(k+1)} a_k, \; k > 1 \qquad (49)$$

We assume that a_0 is equal to 1. Since eq.(49) involves the coefficients $\{a_k\}$, two at a time, we will have two linearly independent solutions involving the even and odd powers of x involving even and odd coefficients $\{a_{2k}\}$ and $\{a_{2k+1}\}$. The even coefficients $\{a_{2k}\}$ for n = 2m are,

$$a_0 = 1$$

$$a_2 = -\frac{1}{2!} n(n+1)$$

$$a_4 = \frac{1}{4!} n(n-2)(n+1)(n+3)$$

$$a_6 = -\frac{1}{6!} n(n-2)(n-4)(n+1)(n+3)(n+5)$$

$$a_8 = \frac{1}{8!} n(n-2)(n-4)(n-6) \times (n+1)(n+3)(n+5)(n+7)$$

$$\vdots$$

$$a_{2k} = -1^k \frac{1}{2k!} n(n-2)(n-4)\ldots(n-2k+2) \times (n+1)(n+3)\ldots(n+2k-1) \qquad (50)$$

$$\vdots$$

$$a_{2m} = -1^m \frac{1}{2m!} 2m(2m-2)(2m-4)\ldots 2 \times (2m+1)(2m+3)\ldots(4m-1)$$

The odd coefficients $\{a_{2k+1}\}$ for n = 2m + 1 are,

$$a_1 = 1$$

$$a_3 = -\frac{1}{3!}(n-1)(n+2)$$

$$a_5 = \frac{1}{5!}(n-1)(n-3)(n+2)(n+4)$$

$$a_7 = -\frac{1}{7!}(n-1)(n-3)(n-5)$$
$$\times (n+2)(n+4)(n+6)$$

$$a_9 = \frac{1}{9!}(n-1)(n-3)(n-5)(n-7)$$
$$\times (n+2)(n+4)(n+6)(n+8)$$

$$\vdots$$

$$a_{2k+1} = -1^k \frac{1}{(2k+1)!}(n-1)(n-3)\ldots(n-2k+1) \times (n+2)(n+4)\ldots(n+2k) \qquad (51)$$

$$\vdots$$

$$a_{2m+1} = -1^m \frac{1}{(2m+1)!} 2m(2m-2)\ldots 2$$
$$\times (2m+3)(2m+5)\ldots(4m+1)$$

Corresponding to the even and odd coefficients eqs.(50, 51), the two linearly independent solutions $f_{even}(x)$ and $f_{odd}(x)$ are,

$$f_{even}(x) = 1 + \sum_{k=1}^{m} (-1)^k 2m(2m-2)(2m-4)\ldots(2m-2k+2) \times (2m+1)(2m+3)\ldots(2m+2k-1)\frac{x^{2k}}{2k!}$$

$$f_{odd}(x) = 1 + \sum_{k=1}^{m} (-1)^k 2m(2m-2)\ldots(2m-2k+2) \times (2m+3)(2m+5)\ldots(2m+2k+1)\frac{x^{2k+1}}{(2k+1)!} \qquad (52)$$

In eqs.(52), the polynomials corresponding to m = 0, 2, 4… in $f_{even}(x)$ are given by,

$\{1, 1 - 3x^2, 1 - 10x^2 + \frac{35}{3}x^4, \cdots\}$

and for m = 1, 3, 5 … in $f_{odd}(x)$ are given by,

$\{x, x - \frac{5}{3}x^3, x - \frac{14}{3}x^3 + \frac{21}{5}x^5 \ldots\}$.

These polynomials, when standardized by $f_{even}(1)$ and $f_{odd}(1)$, yield the Legendre polynomials $\{P_n(x)\}$ given by,

$$P_0(x) = 1$$

$$P_1(x) = x$$

$$P_2(x) = \frac{3}{2}\left(x^2 - 1\right)$$

$$P_3(x) = \frac{1}{2}\left(5x^3 - 3x\right)$$

$$P_4(x) = \frac{1}{8}\left(35x^4 - 30x^2 + 3\right)$$

$$P_5(x) = \frac{1}{8}\left(63x^5 - 70x^3 + 15x\right) \qquad (53)$$

$$P_6(x) = \frac{1}{16}\left(231x^6 - 315x^4 + 105x^2 - 5\right)$$

The above polynomials are not orthogonal, the orthogonality constant being $\sqrt{\frac{2}{2n+1}}$. The normalized Legendre polynomials $P_{n(}x)$ are given by $P_n(x) = \frac{P_n}{\sqrt{\frac{2}{2n+1}}}$. These polynomials from $P_1(x)$ to $P_6(x)$ are shown in the figure below. Fuller details are shown in the next section.

LEGENDRE POLYNOMIALS

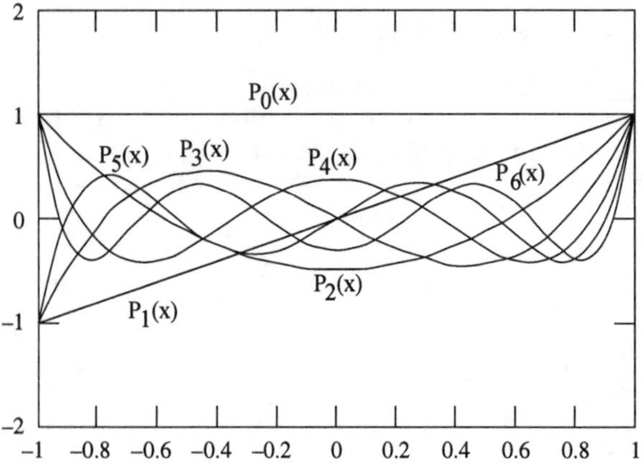

7. USEFUL ORTHOGONAL POLYNOMIALS

CHEBYSEV POLYNOMIALS – FIRST KIND

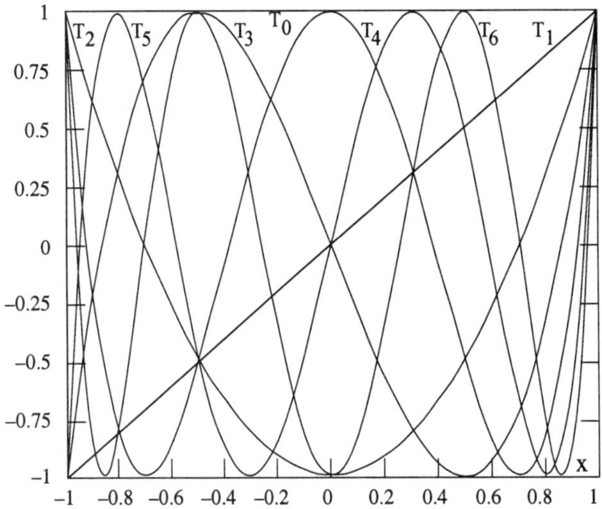

Differential Equation: $\left(1 - x^2\right) \dfrac{d^2y}{dx^2} - x \dfrac{dy}{dx} + n^2\, y = 0$

Recurrence Relation: $T_{n+1}(x) = 2x\ T_n(x) - T_{n-1}(x)$

Weighting Function: $\dfrac{1}{\sqrt{1 - x^2}}$

Orthogonality Interval: $[-1, 1]$

Normalization: $\displaystyle\int_{-1}^{+1} \frac{\left[T_n(x)\right]^2}{\sqrt{1 - x^2}}\, dx = \begin{cases} \dfrac{\pi}{2} & n \neq 0 \\ \pi & n = 0 \end{cases}$

Chebyshev Polynomials - First Kind

$T_0(x) = 1$

$T_1(x) = x$

$T_2(x) = 2x^2 - 1$

$T_3(x) = 4x^3 - 3x$

$T_4(x) = 8x^4 - 8x^2 + 1$

$T_5(x) = 16x^5 - 20x^3 + 5x$

$T_6(x) = 32x^6 - 48x^4 + 18x^2 - 1$

$T_7(x) = 64x^7 - 112x^5 + 56x^3 - 7x$

$T_8(x) = 128x^8 - 256x^6 + 160x^4 - 32x^2 + 1$

$T_9(x) = 256x^9 - 576x^7 + 432x^5 - 120x^3 + 9x$

$T_{10}(x) = 512x^{10} - 1280x^8 + 1120x^6 - 400x^4 + 50x^2 - 1$

CHEBYSEV POLYNOMIALS – SECOND KIND

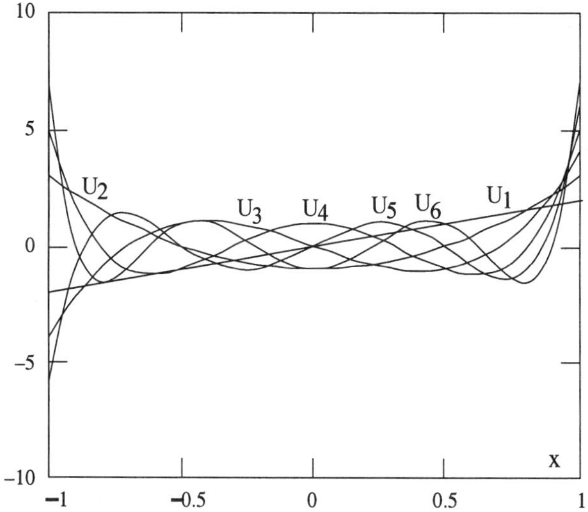

Differential Equation: $\left(1-x^2\right)\dfrac{d^2y}{dx^2} - 3x\,\dfrac{dy}{dx} + n(n+2)y = 0$

Recurrence Relation: $U_{n+1}(x) = 2x\,U_n(x) - U_{n-1}(x)$

Weighting Function: $\sqrt{1 - x^2}$

Orthogonality Interval: $[-1, 1]$

Normalization: $\displaystyle\int_{-1}^{+1} \sqrt{1 - x^2}\,\left[U_n(x)\right]^2 dx = \frac{\pi}{2}$

Chebyshev Polynomials - Second Kind

$U_0(x) = 1$

$U_1(x) = 2x$

$U_2(x) = 4x^2 - 1$

$U_3(x) = 8x^3 - 4x$

$U_4(x) = 16x^4 - 12x^2 + 1$

$U_5(x) = 32x^5 - 32x^3 + 6x$

$U_6(x) = 64x^6 - 80x^4 + 24x^2 - 1$

$U_7(x) = 128x^7 - 192x^5 + 80x^3 - 8x$

$U_8(x) = 256x^8 - 448x^6 + 240x^4 - 40x^2 + 1$

$U_9(x) = 256x^8 - 448x^6 + 240x^4 - 40x^2 + 1$

$U_{10}(x) = 1024x^{10} - 2304x^8 + 1792x^6 - 560x^4 + 60x^2 - 1$

HERMITE POLYNOMIALS

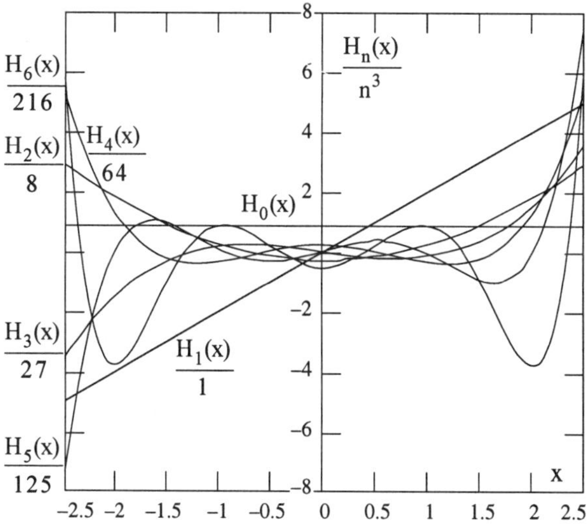

Differential Equation: $$\frac{d^2y}{dx^2} - 2x\frac{dy}{dx} + 2ny = 0$$

Recurrence Relation: $$H_{n+1}(x) = 2\,x\,H_n(x) - 2n\,H_{n-1}(x)$$

Weighting Function: $$e^{-x^2}$$

Orthogonality Interval: $$[-\infty, \infty]$$

Normalization: $$\int_{-\infty}^{\infty} e^{-x^2}\left[H_n(x)\right]^2 dx = 2^n\, n!\,\sqrt{\pi}$$

Hermite Polynomials

$H_0(x) = 1$
$H_1(x) = 2x$
$H_2(x) = 4x^2 - 2$
$H_3(x) = 8x^3 - 12x$
$H_4(x) = 16x^4 - 48x^2 + 12$
$H_5(x) = 32x^5 - 160x^3 + 120x$
$H_6(x) = 64x^6 - 480x^4 + 720x^2 - 120$
$H_7(x) = 128x^7 - 1344x^5 + 3360x^3 - 1680x$
$H_8(x) = 256x^8 - 3584x^6 + 13440x^4 - 13440x^2 + 1680$
$H_9(x) = 512x^9 - 9216x^7 + 48384x^5 - 80640x^3 + 30240x$
$H_{10}(x) = 1024x^{10} - 23040x^8 + 161280x^6 - 403200x^4 + 302400x^2 - 30240$

LAGUERRE POLYNOMIALS

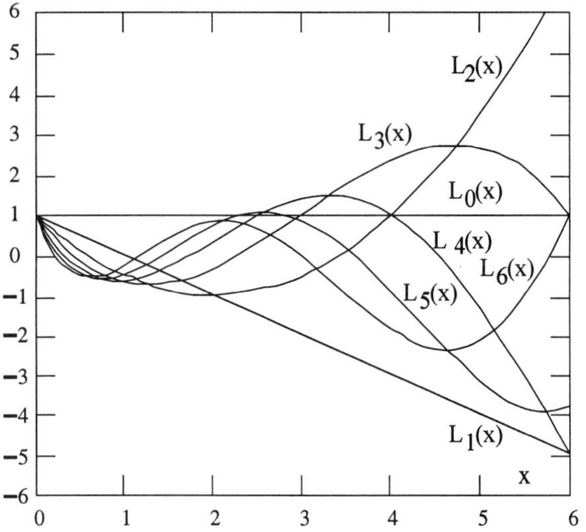

Differential Equation: $$x \frac{d^2y}{dx^2} - (1 - x)\frac{dy}{dx} + ny = 0$$

Recurrence Relation: $$(n+1)\, L_{n+1}(x) = (2n+1-x)\, L_n(x) - n\, L_{n-1}(x)$$

Weighting Function: $$e^{-x}\, u(x)$$

Orthogonality Interval: $$[\,0, \infty)$$

Normalization: $$\int_0^\infty e^{-x}\left[L_n(x)\right]^2 dx = 1$$

Laguerre Polynomials

$L_0(x) = 1$

$L_1(x) = 1 - x$

$L_2(x) = 1 - 2x + \frac{1}{2} x^2$

$L_3(x) = 1 - 3x + \frac{3}{2} x^2 - \frac{1}{6} x^3$

$L_4(x) = 1 - 4x + 3 x^2 - \frac{2}{3} x^3 + \frac{1}{24} x^4$

$L_5(x) = 1 - 5x + 5 x^2 - \frac{5}{3} x^3 + \frac{5}{24} x^4 - \frac{1}{120} x^5$

$$L_6(x) = 1 - 6x + \frac{15}{2} x^2 - \frac{10}{3} x^3 + \frac{5}{8} x^4 - \frac{1}{20} x^5 + \frac{1}{720} x^6$$

$$L_7(x) = 1 - 7x + \frac{21}{2} x^2 - \frac{35}{6} x^3 + \frac{35}{24} x^4 - \frac{7}{40} x^5 + \frac{7}{720} x^6 - \frac{1}{5040} x^7$$

$$L_8(x) = 1 - 8x + 14 x^2 - \frac{28}{3} x^3 + \frac{35}{12} x^4 - \frac{7}{15} x^5 + \frac{7}{180} x^6 - \frac{1}{630} x^7 + \frac{1}{40320} x^8$$

$$L_9(x) = 1 - 9x + 18 x^2 - 14 x^3 + \frac{21}{4} x^4 - \frac{21}{10} x^5 + \frac{7}{60} x^6 - \frac{1}{140} x^7 + \frac{1}{4480} x^8 - \frac{1}{362880} x^9$$

$$L_{10}(x) = 1 - 10x + \frac{45}{2} x^2 - 20 x^3 + \frac{35}{4} x^4 - \frac{21}{10} x^5 + \frac{7}{24} x^6 - \frac{1}{42} x^7 + \frac{1}{896} x^8 - \frac{1}{36288} x^9 + \frac{1}{3628800} x^{10}$$

GENERALIZED LAGUERRE POLYNOMIALS

Differential Equation: $$x\frac{d^2y}{dx^2} - (\alpha + 1 - x)\frac{dy}{dx} + ny = 0$$

Recurrence Relation: $$\begin{aligned}(n+1)\, L_{n+1}(\alpha, x) &= (2n+\alpha+1-x)\, L_n(\alpha, x) \\ &\quad - (n+\alpha)\, L_{n-1}(\alpha, x)\end{aligned}$$

Weighting Function: $x^{\alpha} e^{-x} u(x)$

Orthogonality Interval: $[\,0, \infty)$

Normalization: $$\int_0^{\infty} x^{\alpha} e^{-x} [L_n(x)]^2\, dx = \frac{\Gamma(n + \alpha + 1)}{n!}$$

Generalized Laguerre Polynomials

$$L_0(\alpha, x) = 1$$

$$L_1(\alpha, x) = (1 + \alpha) - x$$

$$L_2(\alpha, x) = \left(1 + \frac{3}{2}\alpha + \alpha^2\right) - (2 + \alpha)\,x + \frac{1}{2}x^2$$

$$L_3(\alpha, x) = \left(1 + \frac{11}{6}\alpha + \alpha^2 + \frac{1}{6}\alpha^3\right) - \left(3 + \frac{5}{2}\alpha + \frac{1}{2}\alpha^2\right)x + \left(\frac{3}{2} + \frac{1}{2}\alpha\right)x^2 - \frac{1}{6}x^3$$

$$L_4(\alpha, x) = \left(1 + \frac{25}{12}\alpha + \frac{35}{24}\alpha^2 + \frac{5}{12}\alpha^3 + \frac{1}{24}\alpha^4\right) - \left(4 + \frac{13}{3}\alpha + \frac{3}{2}\alpha^2 + \frac{1}{6}\alpha^3\right)x + \left(3 + \frac{7}{4}\alpha + \frac{1}{4}\alpha^2\right)x^2 - \left(\frac{2}{3} + \frac{1}{6}\alpha\right)x^3 + \frac{1}{24}x^4$$

$$L_5(\alpha, x) = \left(1 + \frac{137}{60}\alpha + \frac{15}{8}\alpha^2 + \frac{17}{24}\alpha^3 + \frac{1}{8}\alpha^4 + \frac{1}{120}\alpha^5\right) - \left(5 + \frac{77}{12}\alpha + \frac{71}{24}\alpha^2 + \frac{7}{12}\alpha^3 + \frac{1}{24}\alpha^4\right)x + \left(5 + \frac{47}{12}\alpha + \alpha^2 + \frac{1}{12}\alpha^3\right)x^2 - \left(\frac{5}{3} + \frac{3}{4}\alpha + \frac{1}{12}\alpha^2\right)x^3 + \left(\frac{5}{24} + \frac{1}{24}\alpha\right)x^4 - \frac{1}{120}x^5$$

GENERALIZED LAGUERRE POLYNOMIALS WITH $\alpha = 1$

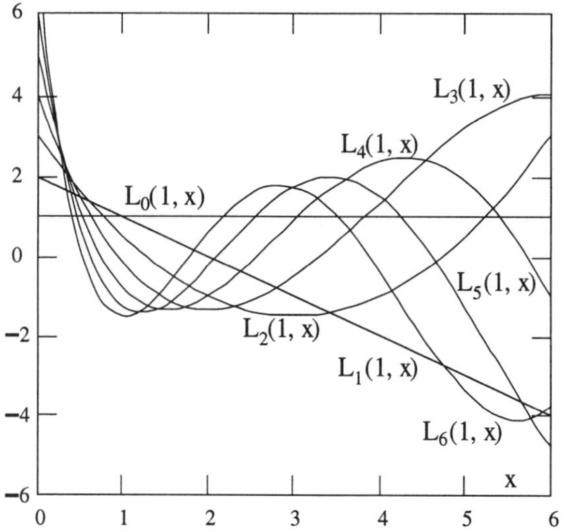

Differential Equation: $$x \frac{d^2y}{dx^2} - (2 - x)\frac{dy}{dx} + ny = 0$$

Recurrence Relation: $$(n+1)\, L_{n+1}(1, x = (2n + 2 - x)\, L_n(1, x) - (n + 1)\, L_{n-1}(1, x)$$

Weighting Function: $xe^{-x}\, u(x)$

Orthogonality Interval: $[0, \infty)$

Normalization: $$\int_0^{\infty} x\, e^{-x} \left[L_n(x)\right]^2 dx = \frac{\Gamma(n+2)}{n!} = n$$

Generalized Laguerre Polynomials with $\alpha = 1$ $[L_0(x) - L_6(x)]$

$L_0(1, x) = 1$

$L_1(1, x) = 2 - x$

$L_2(1, x) = 3 - 3x + \frac{1}{2} x^2$

$L_3(1, x) = 4 - 6x + 2 x^2 - \frac{1}{6} x^3$

$L_4(1, x) = 5 - 10x + 5 x^2 - \frac{5}{6} x^3 + \frac{1}{24} x^4$

$L_5(1, x) = 6 - 15x + 10 x^2 - \frac{5}{2} x^3 + \frac{1}{4} x^4 - \frac{1}{120} x^5$

$L_6(1, x) = 7 - 21x + \frac{35}{2} x^2 - \frac{35}{6} x^3 + \frac{7}{8} x^4 - \frac{7}{20} x^5 + \frac{1}{720} x^6$

LEGENDRE POLYNOMIALS

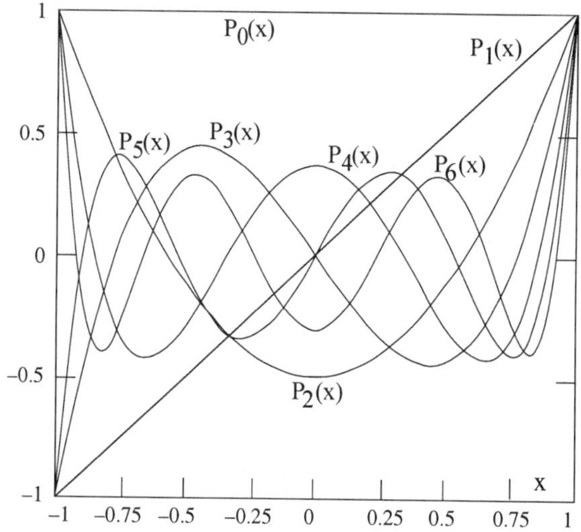

Differential Equation: $(1 - x^2)\dfrac{d^2y}{dx^2} - 2x\dfrac{dy}{dx} + n(n+1)y = 0$

Recurrence Relation: $(n+1)\, P_{n+1}(x) = (2n+1)\, x\, P_n(x) - n\, P_{n-1}(x)$

Weighting Function: 1

Orthogonality Interval: $[-1, 1]$

Normalization: $\int_{-1}^{+1} \left[P_n(x)\right]^2 dx = \dfrac{2}{2n+1}$

Legendre Polynomials $P_0(x) - P_{10}(x)$

$$P_0(x) = 1$$

$$P_1(x) = x$$

$$P_2(x) = \frac{3}{2}\left(x^2-1\right)$$

$$P_3(x) = \frac{1}{2}\left(5x^3-3x\right)$$

$$P_4(x) = \frac{1}{8}\left(35x^4-30x^2+3\right)$$

$$P_5(x) = \frac{1}{8}\left(63x^5-70x^3+15x\right)$$

$$P_6(x) = \frac{1}{16}\left(231x^6-315x^4+105x^2-5\right)$$

$$P_7(x) = \frac{1}{16}\left(429x^7-693x^5+315x^3-35x\right)$$

$$P_8(x) = \frac{1}{128}\left(6435x^8-12012x^6+6930x^4-1260x^2+35\right)$$

$$P_9(x) = \frac{1}{128}\left(12155x^9-25740x^7+18018x^5-4620x^3 +315x\right)$$

$$P_{10}(x) = \frac{1}{256}\left(46189x^{10}-109395x^8+90090x^6 -30030x^4+3465x^2-63\right)$$

8. GRAM-SCHMIDT ORTHONORMALIZATION PROCEDURE

The general Gram-Schmidt orthonormalization procedure takes a set of nonorthogonal linearly independent functions $\{f_k(x), k = 0, 1, \ldots, n\}$ and constructs step by step a set of n orthonormal functions $\{\phi_k(x), k = 0, 1, \ldots, n\}$ with respect to a real weight function w(x) over an interval [a, b]. On the other hand if the set $\{f_k(x), k = 0, 1, \ldots, n\}$ is linearly dependent, then G-S procedure will yield an orthonormal set $\{\phi_k(x)\}$ that is of lesser dimension than the set $\{f_k(x)\}$.

If $\{\psi_k(x)\}$ is an orthogonal set over the interval [a, b] with respect to a *real* weighting function w(x), then

$$\int_a^b |\psi_k(x)|^2 w(x)\, dt = \alpha_{kk}^2 \quad k = 0, 1, \ldots, n \tag{1}$$

and the weighted norm of $\psi_k(x)$ with respect to the weight w(x) is defined by,

$$\|\psi_k(x)\| = \left[\int_a^b |\psi_k(x)|^2 w(x)\, dx\right]^{\frac{1}{2}} = \alpha_{kk} \tag{2}$$

We can now construct an orthonormal sequence from eq.(2) by defining $\phi_k(x) = \dfrac{\psi_k(x)}{\alpha_{kk}}$: k = 0, 1, …n, so that,

$$\int_a^b |\phi_k(x)|^2 w(x)\, dx = 1 : \; k = 0, 1, \ldots, n \tag{3}$$

We shall now show the sequence of steps starting from k = 0 in the orthonormalization procedure for transforming $\{f_k(x), k = 0, 1, \ldots, n\}$ into the weighted orthonormal set $\{\phi_k(x), k = 0, 1, \ldots, m\}$ of dimension m with $m \le n$.

Step 0: Initial Arrangement

Rearrange the sequence $\{f_k(x)\}$ in some suitable order. This rearrangement is an art. Usually this step can be skipped.

Step 1 : k = 0

We first set the first function $f_0(x)$ to $\psi_0(x)$

$$\psi_0(x) = f_0(x) \tag{4}$$

Normalize $\psi_0(x)$ with respect to the weighting function w(x) so that from eq.(2)

$$\phi_0(x) = \frac{\psi_0(x)}{||\psi_0(x)||} \quad \text{or,} \quad \psi_0(x) = \alpha_{00}\,\phi_0(x) \tag{5}$$

and write

$$f_0(x) = \alpha_{00}\phi_0(x) \tag{6}$$

Step 2 : k = 1

We determine $\psi_1(x)$ from $f_1(x)$ by defining

$$\psi_1(x) = f_1(x) - \alpha_{10}\,\phi_0(x) \tag{7}$$

where α_{10} is the weighted inner product defined by

$$\alpha_{10} = \int_a^b \phi_1(x) w(x) \phi_0^*(x) dx = < f_1(x), w(x)\phi_0(x) > \tag{8}$$

Normalization of $\psi_1(x)$ results in $\phi_1(x)$ as shown below:

$$\phi_1(x) = \frac{\psi_1(x)}{||\psi_1(x)||} \quad \text{or,} \quad \psi_1(x) = \alpha_{11}\,\phi_1(x) \tag{9}$$

and $f_1(x)$ can be written as,

$$f_1(x) = \alpha_{10}\phi_0(x) + \alpha_{11}\,\phi_1(x) \tag{10}$$

We continue this process until we exhaust all the n functions of $\{f_k(x)\}$ and the final step for k = n is shown below.

Step n + 1: k = n

We now determine $\psi_n(x)$ from $f_n(x)$ by defining

$$\psi_n(x) = f_n(x) - \alpha_{n0}\phi_0(x) + \alpha_{n1}\,\phi_1(x) + \ldots + \alpha_{n\,n-1}\,\phi_{n-1}(x) \tag{11}$$

where α_{nj} are the weighted inner products defined by

$$\begin{aligned}\alpha_{n\,j} &= \int_a^b f_n(x)\,w(x)\,\phi_j^*(x)\,dx \\ &= \langle f_n(x), w(x)\,\phi_j(x)\rangle,\ \ j = 0, 2, \ldots, n-1\end{aligned} \tag{12}$$

If the original sequence $\{f_k(x)\}$ is linearly independent then none of the ψ_ks and hence the ϕ_ks will be zero. On the other hand, if $\{f_k(x)\}$ is linearly dependent then some of the ψ_ks will be zero and the resulting dimension will be less than n, the dimension of $\{f_k(x)\}$. We shall now present two examples of Gram-Schmidt orthonormalization procedure.

Example 1

In this example, starting from the linearly independent polynomial sequence, $\{f_k(x)\} = \{1, x, x^2, \ldots, x^k, \ldots\}$, we will find a set of orthonormal polynomials in the interval $[0, \infty]$ with respect to the weighting function, $w(x) = e^{-x}u(x)$.

Step 1

The functions $\{f_k(x)\}$ are already in the desired order.

Step 2

Set $\psi_0(x) = 1$. Normalizing with respect to $e^{-x}u(x,)$ we have $||\psi_0(x)||^2 = \int_0^{\infty} \psi_0(x)\, e^{-x}\, dx = 1$ and hence

$$\phi_0(x) = \psi_0(x) = 1 \tag{13}$$

Step 3

Set $\psi_1(x) = x - \alpha_{10}.1$ where

$$\alpha_{10} = \int_0^{\infty} x\, e^{-x}\, dx = 1, \text{ and, } \psi_1(x) = x - 1$$

Normalizing $\psi_1(x)$ with respect to $e^{-x}u(x)$, we have

$$||\psi_1(x)||^2 = \int_0^{\infty} (x-1)^2\, e^{-x}\, dx = 1.$$

Thus,

$$\phi_1(x) = \psi_1(x) = x - 1 \tag{14}$$

Step 4

Set $\psi_2(x) = x^2 - \alpha_{20}.1 - \alpha_{21}.(x-1)$ where

$$\alpha_{20} = \int_0^{\infty} x^2\, e^{-x} dx = 2 \text{ and}$$

$$\alpha_{21} = \int_0^{\infty} x^2(x-1)\, e^{-x} dx = 4$$

Hence, $\psi_2(x) = x^2 - 2 - 4.(x-1) = x^2 - 4x + 2$. Normalizing $\psi_2(x)$ with respect to $e^{-x}u(x)$, we have

$$||\psi_2(x)||^2 = \int_0^{\infty} (x^2 - 4x + 2)^2\, e^{-x} dx = 4$$

Therefore,

$$\phi_2(x) = \frac{x^2 - 4x + 2}{2} = \frac{1}{2}x^2 - 2x + 1 \qquad (15)$$

We can continue this process and the resulting polynomials $\{\phi_k(x), k = 0, 1, 2 \ldots\}$ are called the LaGuerre polynomials and they form an orthonormal set in the interval $[0, \infty]$ with respect to the weight function $e^{-x}\,u(x)$.

Example 2

Five signals, $x_1(t)$, $x_2(t)$, $x_3(t)$, $x_4(t)$, $x_5(t)$ are shown in the diagram. We will extract an orthonormal sequence with weighting function $w(x) = 1$ using the Gram-Schmidt orthogonalization process. As discussed previously, the maximum number of orthonormal sequences is $\{\phi_k(t), k = 1, \ldots, 5\}$ corresponding to the five signals.

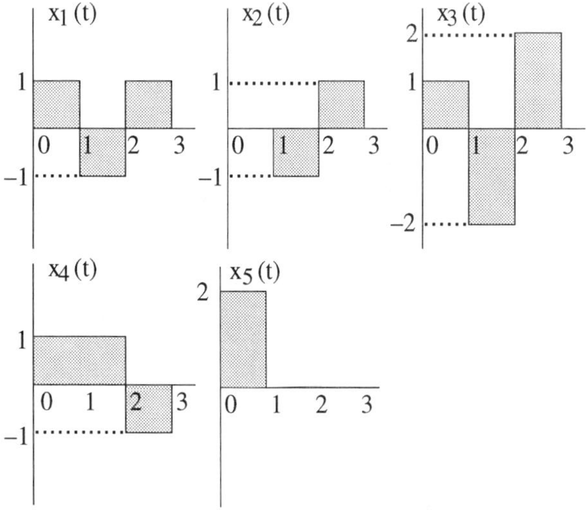

Step 0

$x_1(t)$, $x_2(t)$, $x_3(t)$, $x_4(t)$, $x_5(t)$ have been arranged in a suitable order

Step 1

Set $\psi_1(t) = x_1(t)$. Find

$$E_{\psi 1} = \|\psi_1\|^2 = \int_0^3 \psi_1(t)\, \psi_1^*(t)\, dt$$

Hence $\phi_1(t) = \dfrac{\psi_1(t)}{\sqrt{E_{\psi 1}}} = \dfrac{x_1(t)}{\sqrt{3}}$ and,

$$x_1(t) = \sqrt{3}\, \phi_1(t) + 0.\phi_2(t) + 0.\phi_3(t) + 0.\phi_4(t) + 0.\phi_5(t)$$

$$\mathbf{x}_1 = (\alpha_{11}, 0, 0, 0, 0) = (\sqrt{3}, 0, 0, 0, 0)$$

Step 2

Set $\psi_2(t) = x_2(t) - \langle x_2(t), \phi_1(t)\rangle\, \phi_1(t)$

$$\alpha_{21} = \int_0^3 x_2(t)\, \phi_1^*(t)\, dt = \int_0^3 x_2(t)\, \frac{x_1(t)}{\sqrt{3}}\, dt = \frac{5}{\sqrt{3}}$$

$$\psi_2(t) = x_2(t) - \frac{2}{\sqrt{3}} . \frac{x_1(t)}{\sqrt{3}} = x_2(t) - \frac{2}{3}\, x_1(t)$$

$$\alpha_{22} = \|\, \psi_2(t)\, \| = \sqrt{\frac{2}{3}}$$

Hence $\phi_2(t) = \sqrt{\dfrac{3}{2}}\, \psi_2(t) = \sqrt{\dfrac{3}{2}}\left(x_2(t) - \dfrac{2}{3}\, x_1(t)\right)$

$$\mathbf{x}_2 = (\alpha_{21}, \alpha_{22}, 0, 0, 0) = \left(\frac{5}{\sqrt{3}}, \sqrt{\frac{2}{3}}, 0, 0, 0\right)$$

The orthonormal functions $\phi_1(t)$ and $\phi_2(t)$ are shown below.

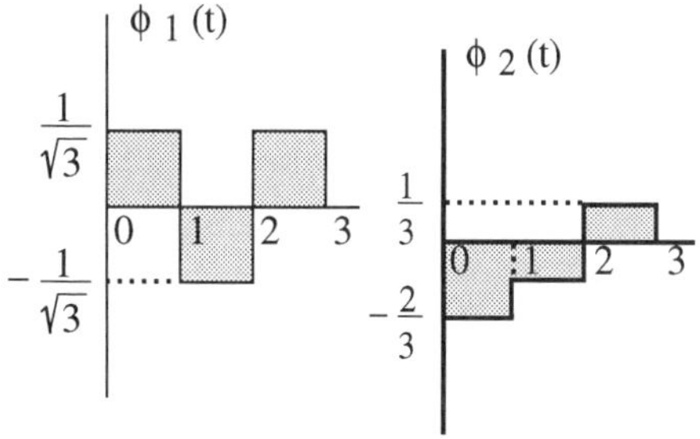

Step 3

Set

$$\psi_3(t) = x_3(t) - \langle x_3(t), \phi_1(t)\rangle\, \phi_1(t) - \langle x_3(t), \phi_2(t)\rangle\, \phi_2(t)$$

$$\alpha_{31} = \langle x_3(t), \phi_1(t)\rangle = \frac{5}{\sqrt{3}}$$

$$\alpha_{32} = \langle x_3(t), \phi_2(t)\rangle = \sqrt{\frac{2}{3}}$$

$$\psi_3(t) = x_3(t) - \frac{5}{\sqrt{3}}\phi_1(t) - \sqrt{\frac{2}{3}}\phi_2(t) = 0.$$

Since $\psi_3(t) = 0$ the dimension becomes one less. Thus,

$$\mathbf{x}_3 = (\alpha_{31}, \alpha_{32}, 0, 0) = \left(\frac{5}{\sqrt{3}}, \sqrt{\frac{2}{3}}, 0, 0\right)$$

Step 4.

Set

$$\psi_4(t) = x_4(t) - \langle x_4(t), \phi_1(t)\rangle\, \phi_1(t) - \langle x_4(t), \phi_2(t)\rangle\, \phi_2(t)$$

$$\alpha_{41} = \langle x_4(t), \phi_1(t)\rangle = -\frac{1}{\sqrt{3}}$$

$$\alpha_{42} = \langle x_4(t), \phi_2(t) \rangle = -2\sqrt{\frac{2}{3}}$$

$$\psi_4(t) = x_4(t) + \frac{1}{\sqrt{3}}\phi_1(t) + 2\sqrt{\frac{2}{3}}\phi_2(t) = 0.$$

Since $\psi_4(t) = 0$ the dimension again becomes one less. Thus,

$$\mathbf{x}_4 = (\alpha_{41}, \alpha_{42}, 0) = \left(-\frac{1}{\sqrt{3}}, -2\sqrt{\frac{2}{3}}, 0\right)$$

Step 5

Set

$$\psi_5(t) = x_5(t) - \langle x_5(t), \phi_1(t) \rangle\, \phi_1(t) - \langle x_5(t), \phi_2(t) \rangle\, \phi_2(t)$$

$$\alpha_{51} = \langle x_5(t), \phi_1(t) \rangle = \frac{2}{\sqrt{3}}$$

$$\alpha_{52} = \langle x_5(t), \phi_2(t) \rangle = -2\sqrt{\frac{2}{3}}$$

$$\psi_5(t) = x_5(t) - \frac{2}{\sqrt{3}}\phi_1(t) + 2\sqrt{\frac{2}{3}}\phi_2(t) = 0.$$

Since $\psi_5(t) = 0$ the dimension again becomes one less. Thus,

$$\mathbf{x}_5 = (\alpha_{51}, \alpha_{52}) = \left(\frac{2}{\sqrt{3}}, -2\sqrt{\frac{2}{3}}\right)$$

The vectors corresponding to $x_1(t)$, $x_2(t)$, $x_3(t)$, $x_4(t)$, $x_5(t)$ in the 2-dimensional function space spanned by the basis vectors $\phi_1(t)$ and $\phi_2(t)$ are shown in the figure on the next page.

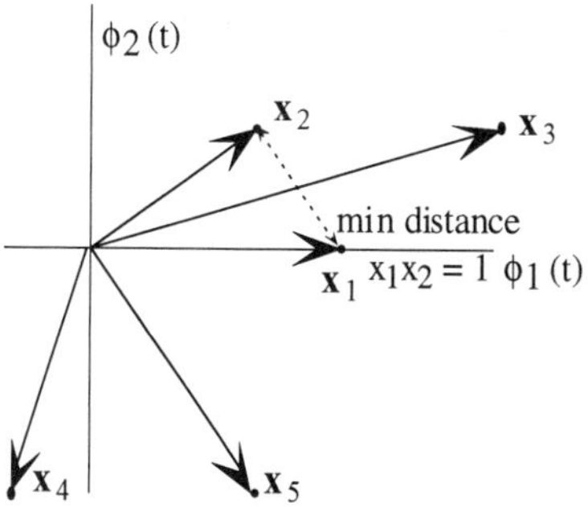

The coordinates corresponding to the vectors $\mathbf{x}_1$, $\mathbf{x}_2$, $\mathbf{x}_3$, $\mathbf{x}_4$, $\mathbf{x}_5$ and the distances between vectors are shown below.

$$\mathbf{x}_1 = \left(\sqrt{3}, 0\right) \qquad x_2 = \left(\frac{2}{\sqrt{3}}, \sqrt{\frac{2}{3}}\right)$$

$$\mathbf{x}_3 = \left(\frac{5}{\sqrt{3}}, \sqrt{\frac{2}{3}}\right) \qquad \mathbf{x}_4 = \left(-\frac{1}{\sqrt{3}}, -2\sqrt{\frac{2}{3}}\right)$$

$$\mathbf{x}_5 = \left(\frac{2}{\sqrt{3}}, -2\sqrt{\frac{2}{3}}\right)$$

$$\overline{\mathbf{x}_1\,\mathbf{x}_2} = 1$$

$$\overline{\mathbf{x}_1\,\mathbf{x}_3} = \sqrt{2} \quad \overline{\mathbf{x}_2\,\mathbf{x}_3} = \sqrt{3}$$

$$\overline{\mathbf{x}_1\,\mathbf{x}_4} = 2\sqrt{2} \quad \overline{\mathbf{x}_2\,\mathbf{x}_4} = 3 \qquad \overline{\mathbf{x}_3\,\mathbf{x}_4} = 3\sqrt{2}$$

$$\overline{\mathbf{x}_1\,\mathbf{x}_5} = \sqrt{3} \quad \overline{\mathbf{x}_2\,\mathbf{x}_5} = \sqrt{6} \quad \overline{\mathbf{x}_3\,\mathbf{x}_5} = 3 \qquad \overline{\mathbf{x}_4\,\mathbf{x}_5} = \sqrt{3}$$

From the distances shown above we see that the minimum distance is between the vectors $\mathbf{x}_1$ and $\mathbf{x}_2$ and is equal to 1.

SUMMARY OF GRAM-SCHMIDT ORTHONORMALIZATION PROCEDURE
Weighting Function w(x) = 1

The Gram-Schmidt orthonormalization procedure takes a set of nonorthogonal linearly independent or dependent set of functions $\{f_k(x), k = 1, \ldots, n\}$ and constructs step by step a set of $m \leq n$ orthonormal functions $\{\phi_k(x), k = 1, \ldots, m\}$ with respect to a weight function w(x) over any interval [a, b]. If the set $\{\psi_k(x)\}$ is an orthogonal set with respect to a weight function w(x), then the integral can be written as,

$$\int_a^b \psi_k^2(x)\, w(x)\, dx = \alpha_{kk}^2 \qquad k = 1, 2, \ldots, n \tag{16}$$

and we can normalize $\psi_k(x)$ by defining $\phi_k(x) = \dfrac{\psi_k(x)}{\alpha_{kk}}$ $k = 1, 2, \ldots,\ n$ so that,

$$\int_a^b \phi_k^2(x)\, w(x)\, dx = 1 \qquad k = 1, 2, \ldots, n \tag{17}$$

We shall now summarize the orthonormalization procedure for weight function w(x) = 1 with the usual definitions of the inner products and norms.

$$\psi_1(x) = f_1(x) : \phi_1(x) = \frac{\psi_1(x)}{||\psi_1(x)||}$$

$$\psi_1(x) = f_2(x) - < f_2(x)\,,\, \phi_1(x) >$$

$$\phi_2(x) = \frac{\psi_2(x)}{||\psi_2(x)||}$$

$$\psi_3(x) = f_3(x) - \langle f_3(x), \phi_1(x) \rangle \phi_1(x) - \langle f_3(x), \phi_2(x) \rangle \phi_2(x)$$

$$\phi_3(x) = \frac{\psi_3(x)}{||\psi_3(x)||}$$

$$\vdots$$

$$\psi_n(x) = f_n(x) - \langle f_n(x), \phi_1(x) \rangle \phi_1(x) - \langle f_n(x), \phi_2(x) \rangle \phi_2(x) - \cdots - \langle f_n(x), \phi_{n-1}(x) \rangle \phi_{n-1}(x) \tag{18}$$

$$\phi_n(x) = \frac{\psi_n(x)}{||\psi_n(x)||}$$

Some of the ϕ_ks can be zero if the original set $\{f_k(x)\}$are linearly dependent. We now define the following generalized Fourier coefficients:

$$\alpha_{kk}^2 = ||\psi_k(x)||^2 = E_{\psi_k} = \int_a^b \psi_k(x)\psi_k^*(x)dx$$

$$\underset{j<k}{\alpha_{kj}} = \langle f_k(x), \phi_j(x) \rangle = \int_a^b \phi_k(x)\phi_j^*(x)dx \tag{19}$$

We can now express the functions $\{f_k(x), k = 1, 2, \ldots, n\}$ in a lower triangular matrix form in terms of the orthonormal bases vectors $\{\phi_k(x), k = 1, 2, \ldots, n\}$ as follows.

$$\begin{bmatrix} f_1(x) \\ f_2(x) \\ \vdots \\ f_k(x) \\ \vdots \\ f_n(x) \end{bmatrix} = \begin{bmatrix} \alpha_{11} & 0 & \cdots & 0 & \cdots & 0 \\ \alpha_{21} & \alpha_{22} & \cdots & 0 & \cdots & 0 \\ \vdots & \vdots & \cdots & \vdots & \cdots & \vdots \\ \alpha_{k1} & \alpha_{k2} & \cdots & \alpha_{kk} & \cdots & 0 \\ \vdots & \vdots & \cdots & \vdots & \cdots & \vdots \\ \alpha_{n1} & \alpha_{n2} & \cdots & \alpha_{nk} & \cdots & \alpha_{nn} \end{bmatrix} \begin{bmatrix} \phi_1(x) \\ \phi_2(x) \\ \vdots \\ \phi_k(x) \\ \vdots \\ \phi_n(x) \end{bmatrix} \tag{20}$$

Example

We shall take a 3-dimensional example and show graphically the steps involved in the Gram-Schmidt procedure. Corresponding to eqs.(18, 20) the 3-dimensional situation can be written as,

$$\begin{bmatrix} \psi_1(x) = f_1(x) \\ \psi_2(x) = f_2(x) - < f_2(x), \phi_1(x) > \phi_1(x) \\ \psi_3(x) = f_3(x) - < f_3(x), \phi_1(x) > \phi_1(x) \\ \qquad - < f_3(x), \phi_2(x) > \phi_2(x) \end{bmatrix} \tag{21}$$

$$\begin{bmatrix} f_1(x) \\ f_2(x) \\ f_3(x) \end{bmatrix} = \begin{bmatrix} \alpha_{11} & 0 & 0 \\ \alpha_{21} & \alpha_{22} & 0 \\ \alpha_{31} & \alpha_{32} & \alpha_{33} \end{bmatrix} \begin{bmatrix} \phi_1(x) \\ \phi_2(x) \\ \phi_3(x) \end{bmatrix} \tag{22}$$

The coefficients $\{\alpha_{11}, \alpha_{22}, \alpha_{33}, \alpha_{21}, \alpha_{31}, \alpha_{32}\}$ in eq.(22) are given by,

$$\alpha_{11}^2 = \int_a^b | \psi_1(x) |^2 \, dx \quad \alpha_{22}^2 = \int_a^b | \psi_2(x) |^2 dx$$

$$\alpha_{33}^2 = \int_a^b | \psi_3(x) |^2 \, dx \quad \alpha_{21} = \int_a^b f_2(x) \, \phi_1^*(x) dx \tag{23}$$

$$\alpha_{31} = \int_a^b f_3(x) \, \phi_1^*(x) \, dx \quad \alpha_{32} = \int_a^b f_3(x) \, \phi_2^*(x) dx$$

The graphical representation of the three-dimensional Gram-Schmidt orthonormalization procedure is shown in the diagram on the next page.

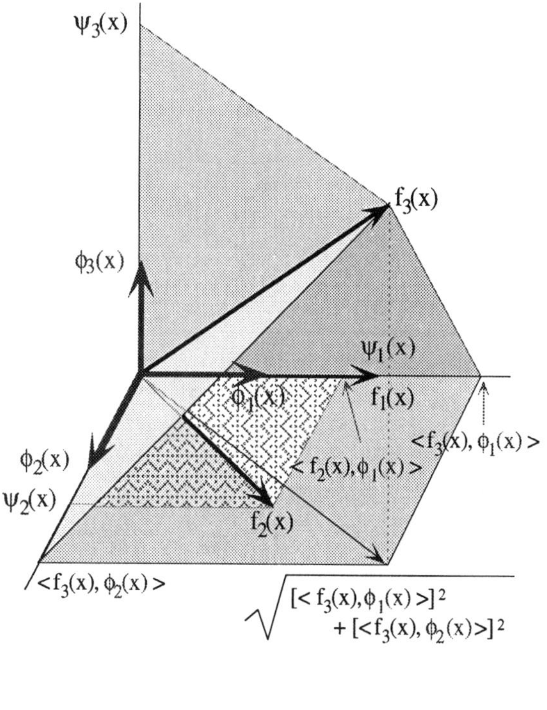
ψ3(x)
f3(x)
φ3(x)
ψ1(x)
φ1(x)
f1(x)
<f3(x), φ1(x) >
φ2(x)
< f2(x), φ1(x) >
ψ2(x)
f2(x)
<f3(x), φ2(x) >
[< f3(x), φ1(x) >]2
+ [<f3(x), φ2(x)>]2

9. PROPERTIES OF CONTINUOUS FOURIER SERIES (CFS)

Definition: $\mathbf{x(t) \Leftrightarrow X_k}$

Periodic Function: $x(t) = x(t + nT_0)$, $n = 0, \pm 1, \ldots$

Fundamental Period: T_0

Fundamental Frequency: $\omega_0 = \dfrac{2\pi}{T_0}$

Exponential Form:

$$x(t) = \sum_{k=-\infty}^{\infty} X_k \, e^{jk\omega_0 t}$$

$$X_k = \frac{1}{T_0} \int_{t_1}^{t_1 + T_0} x(t) \, e^{-jk\omega_0 t} \, dt : \; t_1 \text{ arbitrary}$$

Trigonometric Form:

$$x(t) = a_0 + \sum_{k=1}^{\infty} \left[a_k \cos k\omega_0 t + b_k \sin k\omega_0 t \right]$$

$$a_0 = X_0, \quad \frac{a_k + jb_k}{2} = X_k$$

Phasor Form:

$$x(t) = A_0 + \sum_{k=1}^{\infty} A_k \cos(k\omega_0 t - \theta_k) : a_0 = A_0 = X_0,$$

$$A_k = 2\,|X_k| = \sqrt{a_k^2 + b_k^2}, \; \theta_k = \tan^{-1}\left(b_k/a_k\right)$$

Parseval's Relation:

$$\frac{1}{T_0} \int_{t_1}^{t_1 + T_0} |x(t)|^2 \, dt = \sum_{k=-\infty}^{\infty} |X_k|^2, \; t_1 \text{ arbitrary}$$

Periodic Signal	Property	F Coefficient
GENERAL PROPERTIES		
$x(t) = x(t + nT_0)$	Periodic with period T_0	X_k
$y(t) = y(t + nT_0)$	Periodic with period T_0	Y_k
$ax(t) + by(t)$	Linearity	$aX_k + bY_k$
$x(t - t_0)$	Time shifting	$X_k e^{jk\omega_0 t_0}$
$e^{jm\omega_0 t}\, x(t)$	Frequency shifting	X_{k-m}
$x^*(t)$	Complex conjugate	X^*_{-k}
$x(-t)$	Folding	X_{-k}
$x(at),\ a > 0$	Scaling (periodic with period T_0/α)	X_k
$\int_{t_1}^{t_1+T_0} x(\tau)y(t-\tau)\, d\tau$	Convolution	$T_0 X_k Y_k$
$x(t)y(t)$	Product	$\sum_{m=-\infty}^{\infty} X_m Y_{k-m}$
$\frac{dx}{dt}$	Differentiation	$jk\omega_0 X_k$
$\int_{-\infty}^{t} x(t)\, dt$	Integration (finite valued and periodic only if $X_0 = 0$)	$\frac{X_k}{jk\omega_0}$
SYMMETRY PROPERTIES		
$x(t)$	Real function	$X_k = X^*_{-k}$ $\lvert X_k \rvert = \lvert X_{-k} \rvert$ $-\angle X_k = \angle X_{-k}$

Periodic Signal	Property	F Coefficient
$x(t)$	Real function	$\text{Re}\{X_k\}=\text{Re}\{X_{-k}\}$ $\text{Im}\{X_k\}=\text{Im}\{X_{-k}\}$
$x_e(t) = \text{Ev}\{x(t)\}$	x(t) real	$X_{ek} = a_k$
$x_o(t) = \text{Od}\{x(t)\}$	x(t) real	$X_{ok} = \pm b_k,\ X_{o0} = b_0 = 0$
$x(t) = +x(-t)$	x(t) even	$2X_k = a_k \pm j0$
$x(t) = -x(-t)$	x(t) odd	$2X_k = 0 \pm jb_k$
$x(t) = -x\left(t \pm \frac{T_0}{2}\right)$	Rotation or half-wave odd symmetry	Odd harmonics only
$x(t) = +x\left(t \pm \frac{T_0}{2}\right)$	Half-wave even symmetry. (Used only in a relative sense.) Period $T_1 = T_0/2$	Even harmonics only

TRUNCATED FOURIER SERIES

Gibbs Phenomenon

The sum of a finite number of terms, $x_N(t)$, of a Fourier series is given by

$$x_N(t) = \sum_{k=-N}^{N} X_k\, e^{-jk\omega_0 t}$$

$$= \frac{1}{T_0} \int_{t_1}^{t_1+T_0} x(\tau) \frac{\sin\left[\left(N + \frac{1}{2}\right)\omega_0(t-\tau)\right]}{\sin\left[\omega_0 \frac{(t-\tau)}{2}\right]}\, d\tau$$

$$= \frac{1}{T_0} \int_{t_1}^{t_1+T_0} x(t-\tau) \frac{\sin\left[\left(N+\frac{1}{2}\right)\omega_0\tau\right]}{\sin\left[\omega_0\frac{\tau}{2}\right]} d\tau$$

with t_1 being arbitrary.

The adjoining figures show truncated Fourier series for a periodic pulse of height 1 for N = 5 and N = 10.

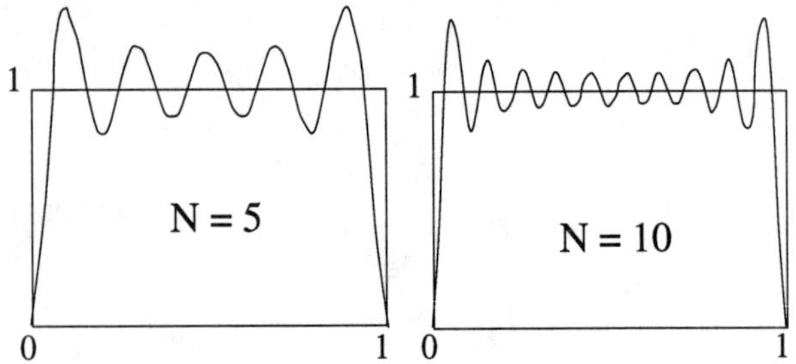

10. FOURIER TRANSFORM AS LIMITING FORM OF FOURIER SERIES

Fourier Coefficient X_k

$$X_k = \frac{2\tau}{T_0} \frac{\sin(k\omega_0\tau)}{k\omega_0\tau}$$

$T_0 = 4\tau : \omega_0 = \frac{2\pi}{4\tau} = \frac{\pi}{2\tau}$

$$T_0 X_k = 2\tau \frac{\sin\left(\frac{k\pi}{2}\right)}{\frac{k\pi}{2}}$$

$T_0 = 8\tau : \omega_0 = \frac{2\pi}{8\tau} = \frac{\pi}{4\tau}$

$$T_0 X_k = 2\tau \frac{\sin\left(k\pi/4\right)}{k\pi/4}$$

$T_0 = 16\tau$

$\omega_0 = \frac{2\pi}{16\tau} = \frac{\pi}{8\tau}$

$$T_0 X_k = 2\tau \frac{\sin\left(k\pi/8\right)}{k\pi/8}$$

$T_0 \rightarrow \infty$

$\omega_0 \rightarrow d\omega : k\omega_0 \rightarrow \omega$

$\lim_{T_0 \rightarrow \infty} T_0 X_k = X(\omega)$

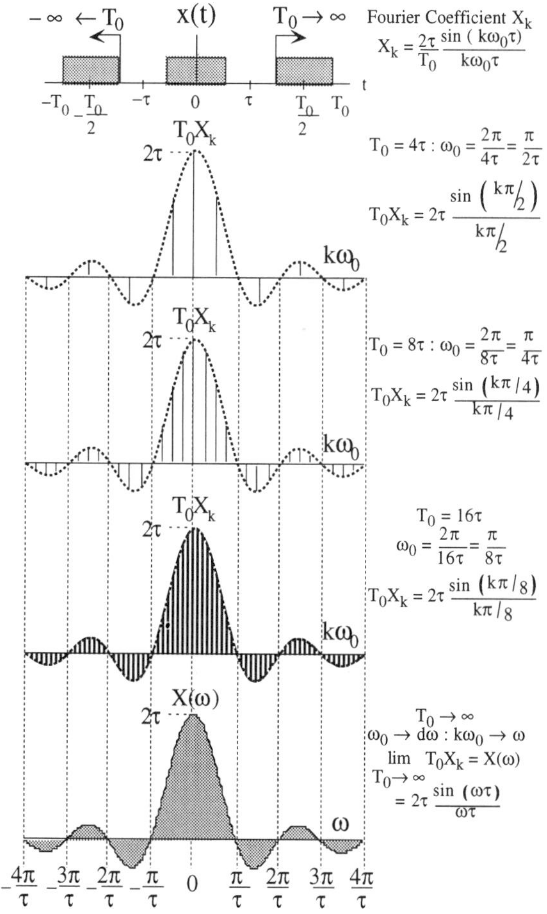

11. PROPERTIES OF CONTINUOUS FOURIER TRANSFORMS (CFT)

DEFINITION: $\mathbf{x(t) \overset{\mathcal{F}}{\longleftrightarrow} X(\omega)}$

Fourier Transform (FT): $X(\omega) = \int_{-\infty}^{\infty} x(t)\, e^{-j\omega t}\, dt$

Inverse FT $x(t) = \frac{1}{2\pi} \int_{-\infty}^{\infty} X(\omega)\, e^{j\omega t}\, d\omega$

FT from FC $X(\omega) = \lim_{T_0 \to \infty} T_0 X_k$

Time Sequence	**Property**	**Fourier Transform**
$a_1x_1(t) + a_2x_2(t)$	Linearity	$a_1X_1(\omega) + a_2X_2(\omega)$
$x(at)$	Scaling	$\frac{1}{\lvert a \rvert} X\left(\frac{\omega}{a}\right)$
$x(-t)$	Folding	$X(-\omega) = X^*(\omega)$
$\int_{-\infty}^{\infty} x(t)\, dt$	Area under curve x(t)	$X(0)$
$2\pi\, x(0)$	Area under curve X(ω)	$\int_{-\infty}^{\infty} X(\omega)\, d\omega$
$X(t)$	Duality	$2\pi\, x(-\omega)$
$x(t \pm t_0)$	Time Shift	$X(\omega)\, e^{\pm j\omega t0}$
$x(t)\, e^{\pm j\omega 0t}$	Frequency Shift Simple Modulation	$X(\omega \mp \omega_0)$

Time Sequence	Property	Fourier Transform
$\frac{d^n x(t)}{dt^n}$	Time Differentiation	$(j\omega)^n X(\omega)$
$(-jt)^n x(t)$	Frequency Differentiation	$\frac{d^n X(\omega)}{d\omega^n}$
$\int_{-\infty}^{t} x(t)\, dt$	Time Integration	$\frac{X(\omega)}{j\omega} + \pi X(0)\delta(\omega)$
$x_1(t) * x_2(t)$	Time Convolution	$X_1(\omega)\, X_2(\omega)$
$x_1(t)x_2(t)$	Frequency Convolution (Modulation)	$\frac{1}{2\pi} X_1(\omega) * X_2(\omega)$
$\int_{-\infty}^{\infty} \lvert x(t)\rvert^2\, dt$	Parseval's Theorem	$\frac{1}{2\pi} \int_{-\infty}^{\infty} \lvert X(\omega)\rvert^2\, d\omega$

12. CONTINUOUS FOURIER TRANSFORM (CFT) PAIRS

Definition: $x(t) \overset{\mathcal{F}}{\longleftrightarrow} X(\omega)$

Fourier Transform (FT): $$X(\omega) = \int_{-\infty}^{\infty} x(t)\, e^{-j\omega t}\, dt$$

Inverse FT $$x(t) = \frac{1}{2\pi} \int_{-\infty}^{\infty} X(\omega)\, e^{j\omega t}\, d\omega$$

Time Function	Fourier Transform
Energy Signals	
1. $p_t(t) = [u(t+t) - u(t-t)]$	$2t\,\frac{\sin \omega t}{\omega t} = 2t\,Sa(\omega t)$
2. $\frac{\omega_0}{\pi}\,\frac{\sin \omega_0 t}{\omega_0 t} = \frac{\omega_0}{\pi}\,Sa(\omega_0 t)$	$p_{\omega 0}(\omega)$
3. $q_\tau(t) = 1 - \frac{\lvert t \rvert}{\tau},\ \lvert t \rvert < \tau$	$\tau \left(\frac{\sin \frac{\omega\tau}{2}}{\frac{\omega\tau}{2}} \right)^2 = \tau Sa^2\left(\frac{\omega\tau}{2}\right)$
4. $\frac{\omega_0}{2\pi} \left(\frac{\sin \frac{\omega_0 t}{2}}{\frac{\omega_0 t}{2}} \right)^2 = \frac{\omega_0}{2\pi}\,Sa^2\left(\frac{\omega_0 t}{2}\right)$	$q_{\omega 0}(\omega)$
5. $\frac{1}{\tau\sqrt{2\pi}}\, e^{-\frac{1}{2}\left(\frac{t}{\tau}\right)^2}$	$e^{-\frac{1}{2}(t\,\omega)^2}$

Time Function	Fourier Function
6. $e^{-at}\,u(t)$	$\dfrac{1}{\alpha + j\omega}$
7. $e^{at}\,u(-t)$	$\dfrac{1}{\alpha - j\omega}$
8. $t\,e^{-at}\,u(t)$	$\dfrac{1}{(\alpha + j\omega)^2}$
9. $e^{-\alpha\|t\|}$	$\dfrac{2\alpha}{\alpha^2 + \omega^2}$
10 $\|t\|e^{-\alpha\|t\|}$	$\dfrac{2(\alpha^2 - \omega^2)}{(\alpha^2 + \omega^2)^2}$
11. $\dfrac{\alpha}{\pi(\alpha^2 + t^2)}$	$e^{-\alpha\|\omega\|}$
12. $e^{-\alpha t}\sin\beta t\,u(t)$	$\dfrac{\beta}{(\alpha^2 + \beta^2) - \omega^2 + j\,2\alpha\omega}$
13. $e^{-\alpha t}\cos\beta t\,u(t)$	$\dfrac{\alpha + j\omega}{(\alpha^2 + \beta^2) - \omega^2 + j\,2\alpha\omega}$

Power and Singularity Signals

14. sgn t	$\dfrac{2}{j\omega}$
15. $\dfrac{1}{jt}$ (Duality)	$-\pi\,\mathrm{sgn}\,\omega$
16. $\delta(t)$ (Dirac delta function)	1
17. δ_n (Kronecker delta function)	$2\pi\,\delta(\omega)$

Time Function	Fourier Function
18. $u(t) = \frac{1}{2} + \frac{1}{2}\ \text{sgn}\ t$	$\pi\,\delta(\omega) + \frac{1}{j\omega}$
19. $\frac{d^n\delta(t)}{dt^n} = \delta^{(n)}(t)$	$(j\omega)^n$
20. $e^{\pm j\omega 0t}$	$2\pi\,\delta(\omega \mp \omega_0)$
21. t^n	$2\pi\, j^n\, \delta^{(n)}(\omega)$
22. $\cos\omega_0 t$	$\pi[\delta(\omega + \omega_0) + \delta(\omega - \omega_0)]$
23. $\sin\omega_0 t$	$j\pi[\delta(\omega + \omega_0) - \delta(\omega - \omega_0)]$
24. $\cos\omega_0 t\ .u(t)$	$\frac{\pi}{2}\left[\delta(\omega+\omega_0) + \delta(\omega-\omega_0)\right] + \frac{j\omega}{\omega_0^2 - \omega^2}$
25. $\sin\omega_0 t\ .u(t)$	$\frac{\pi}{2}\left[\delta(\omega+\omega_0) - \delta(\omega-\omega_0)\right] + \frac{\omega_0}{\omega_0^2 - \omega^2}$
26. $\sum_{n=-\infty}^{\infty} \delta(t - nT_0)$ (comb function)	$\omega_0 \sum_{n=-\infty}^{\infty} \delta(\omega - n\omega_0) : \omega_0 = \frac{2\pi}{T_0}$ (comb function)
27. $x(t) = x(t + mT_0) = \sum_{k=-\infty}^{\infty} X_k\, e^{jk\omega_0 t}$	$2\pi \sum_{k=-\infty}^{\infty} X_k\, \delta(\omega - k\omega_0)$ $\omega_0 = \frac{2\pi}{T_0}$

Definitions of Rectangular and Triangular Pulse Functions

We will now define some of the functions encountered in the formulation of the CFT pairs. The rectangular and triangular pulse functions $p_\tau(t)$ and $q_\tau(t)$ are defined as follows:

$$p_\tau(t) = \begin{cases} 1, & |t| < \tau \\ 0, & \text{otherwise} \end{cases} \tag{1}$$

$$q_\tau(t) = \begin{cases} 1 - \dfrac{|t|}{\tau}, & |t| < \tau \\ 0, & \text{otherwise} \end{cases} \tag{2}$$

These functions are shown in diagrams below:

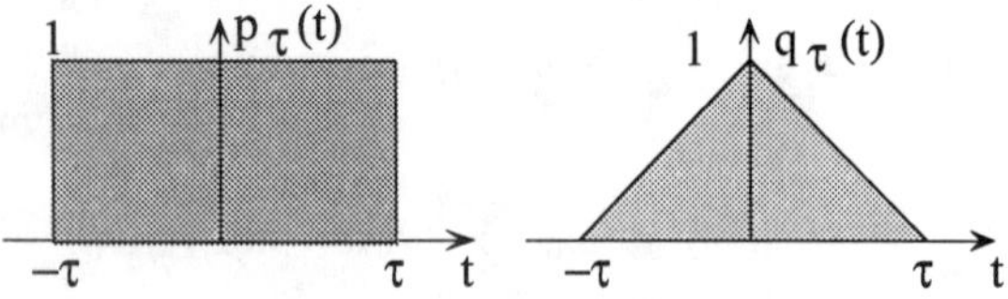

The Sa(x) and sinc(x) are defined by,

$$\mathrm{Sa}(x) = \frac{\sin(x)}{x} \; : \; \mathrm{sinc}(x) = \frac{\sin(\pi x)}{\pi x} \tag{3}$$

and the signum function sgn(x) is defined by,

$$\mathrm{sgn}(x) = \begin{cases} 1, & x > 0 \\ -1, & x < 0 \end{cases} \tag{4}$$

13. INVERSE FOURIER TRANSFORM BY CONTOUR INTEGRATION

(Representation in complex λ-plane)

Complex s-plane: $s = \sigma + j\omega$: identified as Laplace transform or the imaginary frequency plane.

Complex λ-plane: $\lambda = \omega + j\sigma$: identified as Fourier transform or the real frequency plane

Complex λ-plane is a rotation of the complex s-plane by –90 degrees followed by complex conjugation, that is, $[-j(\sigma + j\omega)]^* = \omega + j\sigma$ or $\lambda = (-js)^* = js^*$ and $s = j\lambda^*$. The mapping of the s-plane with poles to the λ-plane with poles is shown in the diagram below.

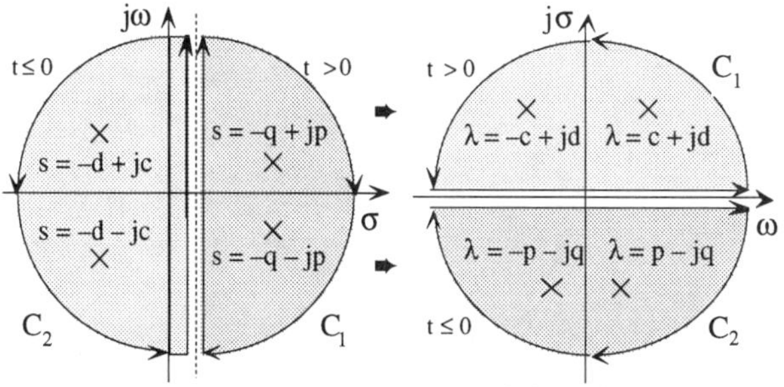

s-plane **λ-plane**

In the complex λ-plane diagram, the contour C_1 in the upper half-plane is traversed in the counter-clockwise direction (*positive contour*) and the integral evaluated along this contour will be the *positive* sum of residues. On the other hand, the contour C_2 being traversed in the clockwise direction (*negative contour*), the integral evaluated along this contour will be the *negative* sum of residues.

The inverse Fourier transform as a contour integral in the real frequency plane is given by:

$$\begin{aligned}
\text{For } t > 0:\ x(t) &= \frac{1}{2\pi}\oint_{C_1} X(\lambda)\, e^{j\lambda t}\, d\lambda \\
&= +j\sum \text{Residues in } C_1 : \text{(CCW)} \\
\text{For } t \le 0:\ x(t) &= \frac{1}{2\pi}\oint_{C_2} X(\lambda)\, e^{j\lambda t}\, d\lambda \\
&= -j\sum \text{Residues in } C_2 : \text{(CW)}
\end{aligned} \tag{1}$$

where $X(\lambda)$ is related to $X(\omega)$ by replacing ω by λ. For the example in the pole-zero constellation of the following figure,

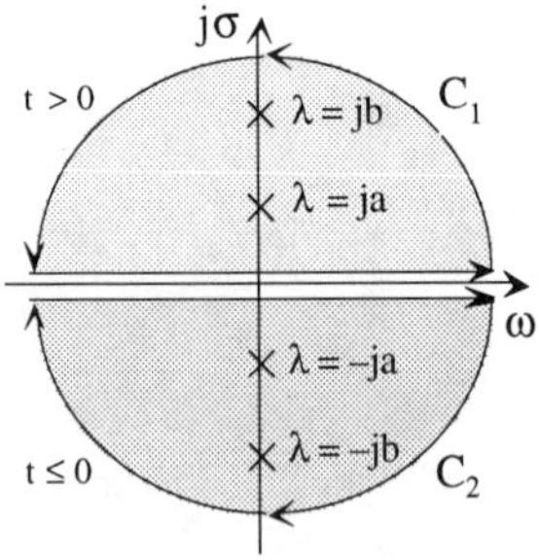

one possible representation of $X(\lambda)$ may be:

$$X(\lambda) = \frac{1}{j}\left(\frac{1}{\lambda - ja} + \frac{1}{\lambda - jb} - \frac{1}{\lambda + ja} - \frac{1}{\lambda + jb}\right) \tag{2}$$

For the representation of $X(\lambda)$ in eq.(2) the transfer function $X(\omega)$ is given by:

$$X(\omega) = \frac{2a}{\omega^2 + a^2} + \frac{2b}{\omega^2 + b^2} \tag{3}$$

and the inverse Fourier transform corresponding to eq.(3) is obtained from the inversion integral given by eq.(1) as follows:

For t > 0:

$$x(t) = \frac{1}{2\pi} \oint_{C_1} \frac{1}{j} \left[\frac{1}{\lambda - ja} + \frac{1}{\lambda - jb} \right] e^{j\lambda t}\, d\lambda$$

$$= + \left(e^{-at} + e^{-bt} \right) \qquad (4)$$

For t ≤ 0:

$$x(t) = \frac{1}{2\pi} \oint_{C_2} \frac{-1}{j} \left[\frac{1}{\lambda + ja} + \frac{1}{\lambda + jb} \right] e^{j\lambda t}\, d\lambda$$

$$= + \left(e^{+at} + e^{+bt} \right)$$

Thus, the complete time function for all t is given by:

$$x(t) = (e^{-at+}e^{-bt)}u(t)+(e^{at}\ e^{bt)}u(-t) = e^{-a|t|}+e^{-b|t|} \qquad (5)$$

Example

We shall find the inverse Fourier transform of

$$X(\omega) = \frac{1}{\omega^4 + a^4}$$

using contour integration of eq.(1) in the complex λ-plane. We shall assume that a is a positive constant. Using the techniques in the previous section we can write

$$X(\lambda) = \frac{1}{\lambda^4 + a^4}$$

and evaluate the contour integral

$$x(t) = \frac{1}{2\pi} \oint_{C_1} \frac{j}{\lambda^4 + a^4} e^{j\lambda t}\, d\lambda = +j \sum \text{residues} \qquad (6a)$$

for t ≥ 0 and

$$x(t) = \frac{1}{2\pi} \oint_{C_2} \frac{j}{\lambda^4 + a^4} e^{j\lambda t}\, d\lambda = -j \sum \text{residues} \qquad (6b)$$

for $t < 0$. The four symmetrically located poles are shown in the following diagram.

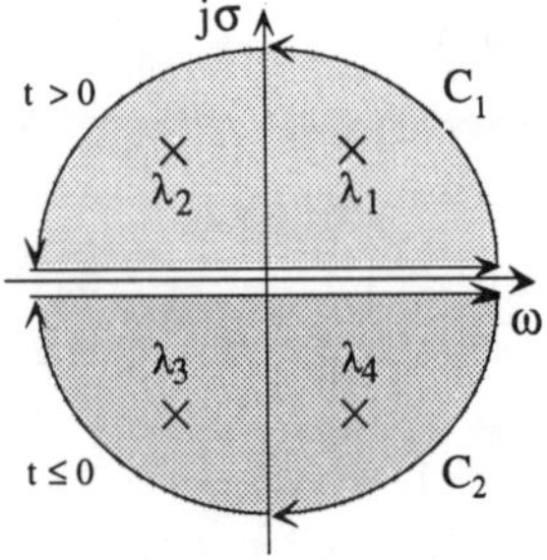

Factoring the denominator polynomial in eqs.(6) we obtain the factors,

$$\lambda_1 = a\left(\frac{1+j}{\sqrt{2}}\right), \quad \lambda_2 = -a\left(\frac{1-j}{\sqrt{2}}\right),$$
$$\lambda_3 = -a\left(\frac{1+j}{\sqrt{2}}\right), \quad \lambda_4 = a\left(\frac{1-j}{\sqrt{2}}\right) \qquad (7)$$

x(t) can now be written as the contour integration around the poles in the positive and negative contours, C_1 and C_2, as shown below:

$$x(t) = \frac{1}{2\pi}\oint_C \frac{e^{j\lambda t}\, d\lambda}{\left[\lambda^2 + ja^2\right]\left[\lambda^2 - ja^2\right]}$$

$$= \frac{1}{2\pi}\oint_{C_1} \frac{e^{j\lambda t}\, d\lambda}{\left[\lambda - a\left(\frac{1+j}{\sqrt{2}}\right)\right]\left[\lambda + a\left(\frac{1-j}{\sqrt{2}}\right)\right]}$$

$$+\frac{1}{2\pi}\oint_{C_2}\left\{\frac{e^{j\lambda t}\,d\lambda}{\left[\lambda+a\left(\frac{1+j}{\sqrt{2}}\right)\right]\left[\lambda-a\left(\frac{1-j}{\sqrt{2}}\right)\right]}\right. \qquad (8)$$

Using contour integration, the residues evaluated at the poles λ_1 and λ_2 in the positive contour C_1 are:

Residue at $\lambda_1 = a\left(\frac{1+j}{\sqrt{2}}\right)$ is given by:

$$\begin{aligned}\mathrm{Res}_{\lambda_1} &= \frac{e^{\left(-a t/\sqrt{2}\right)}\, e^{\left(j a t/\sqrt{2}\right)}}{\left[2ja^2\right]\left[2a\,\frac{(1+j)}{\sqrt{2}}\right]} \\ &= \frac{e^{\left(-a t/\sqrt{2}\right)}\, e^{\left(j a t/\sqrt{2}\right)}\,(1-j)}{j4\sqrt{2}\;a^3}\end{aligned} \qquad (9)$$

Similarly, residue at $\lambda_2 = a\left(\frac{-1+j}{\sqrt{2}}\right)$ is given by:

$$\begin{aligned}\mathrm{Res}_{\lambda_2} &= \frac{e^{\left(-a t/\sqrt{2}\right)}\, e^{\left(-j a t/\sqrt{2}\right)}}{\left[2ja^2\right]\left[2a\,\frac{(1+j)}{\sqrt{2}}\right]} \\ &= \frac{e^{\left(-a t/\sqrt{2}\right)}\, e^{\left(-j a t/\sqrt{2}\right)}\,(1+j)}{j4\sqrt{2}\;a^3}\end{aligned} \qquad (10)$$

Thus, for $t \geq 0$, $x(t)$ is given by j times the positive sum of the residues at λ_1 and λ_2. Or,

$$x(t) = \frac{e^{\left(-a t/\sqrt{2}\right)}}{2\sqrt{2}\;a^3}\left[\cos\left(a\,t/\sqrt{2}\right) - \sin\left(a\,t/\sqrt{2}\right)\right] u(t) \qquad (11)$$

In a similar manner, x(t) for $t \le 0$ can be evaluated as j times the negative sum of residues at λ_3 and λ_4 and the final result for all t is given by:

$$x(t) = \left\{ \frac{e^{\left(-a t/\sqrt{2}\right)}}{2\sqrt{2}\, a^3} \left[\cos\left(a t/\sqrt{2}\right) - \sin\left(a t/\sqrt{2}\right)\right] \right\} u(t)$$
$$+ \left\{ \frac{e^{\left(a t/\sqrt{2}\right)}}{2\sqrt{2}\, a^3} \left[\cos\left(a t/\sqrt{2}\right) + \sin\left(a t/\sqrt{2}\right)\right] \right\} u(-t) \qquad (12)$$

14. DERIVATION OF CONTINUOUS HILBERT TRANSFORMS (CHT)

Definition: $\mathbf{x(t)} \overset{\mathcal{H}}{\longleftrightarrow} \hat{\mathbf{x}}\mathbf{(t)} : \mathbf{X(\omega)} \overset{\mathcal{H}}{\longleftrightarrow} \hat{\mathbf{X}}\mathbf{(\omega)}$

$$\hat{x}(t) = x(t) * \frac{1}{\pi t} = \frac{1}{\pi} \mathcal{P}\left[\int_{-\infty}^{\infty} \frac{x(t)}{t-\tau}\, d\tau\right]$$

$$= \lim_{\varepsilon \to 0} \int_{-\infty}^{-\varepsilon} \frac{x(t)}{t-\tau}\, d\tau + \int_{\varepsilon}^{\infty} \frac{x(t)}{t-\tau}\, d\tau \tag{1}$$

where $\mathcal{P}$ stands for the Cauchy principal value of the improper integral that can be evaluated by contour integration as,

$$\hat{x}(t) = \frac{1}{\pi} \mathcal{P}\left[\int_{-\infty}^{\infty} \frac{x(t)}{t-\tau}\, d\tau\right]$$

$$= \frac{1}{\pi} \oint_{C} \frac{x(z)}{t-z}\, dz = 2j \sum \text{residues of poles in C} \tag{2}$$

where C is a suitably chosen positive contour in the complex z-plane

Assumptions

1. x(t) is square integrable
2. $X(\omega)$ is minimum phase

Definitions

FT of $\hat{x}(t)$: $\hat{x}(t) \overset{\mathcal{F}}{\longleftrightarrow} \hat{X}(\omega) = X(\omega).[-j\, \mathrm{sgn}(\omega)]$ (3)

HT of $X(\omega)$: $X(\omega) \overset{\mathcal{H}}{\longleftrightarrow} \hat{\hat{X}}(\omega) = X(\omega) * \frac{1}{\pi\, \omega}$

$$= \frac{1}{\pi} \mathcal{P}\left[\int_{-\infty}^{\infty} \frac{x(\lambda)}{\omega - \lambda} d\lambda\right] \qquad (4)$$

Note that, in general, $\widehat{X}(\omega) \neq \widehat{\widehat{X}}(\omega)$, but if x(t) is a real causal time function, then $\widehat{X}(\omega) = -\widehat{\widehat{X}}(\omega)\, \text{sgn}(\omega)$. The time function corresponding to $\widehat{\widehat{X}}(\omega)$ is given by,

$$\widehat{\widehat{X}}(\omega) = X(\omega)*\frac{1}{\pi\,\omega} \Leftrightarrow 2\pi\left[x(t).\ \frac{j\text{sgn}(t)}{2\pi}\right] \qquad (5)$$

$$= jx(t)\,\text{sgn}(t) = \hat{\hat{x}}(t) \neq \hat{x}(t)$$

Let x(t) be defined by $x(t) = x^+(t) + x^-(t)$ where $x^+(t)$ is causal and $x^-(t)$ is anticausal. Hence,

$$\hat{\hat{x}}(t) = jx(t)\text{sgn}(t) = j[x^+(t) - x^-(t)]$$

Let $x^+(t) \overset{\mathcal{F}}{\leftrightarrow} X^+(\omega)$ and $x^-(t) \overset{\mathcal{F}}{\leftrightarrow} X^-(\omega)$ and $X(\omega) = X^+(\omega) + X^-(\omega)$. In the complex λ-plane, $X^+(\lambda)$ has poles and zeros in the upper half of the plane (UHP) and $X^-(\lambda)$ has poles and zeros in the lower half of the plane (LHP) as shown in the following diagram.

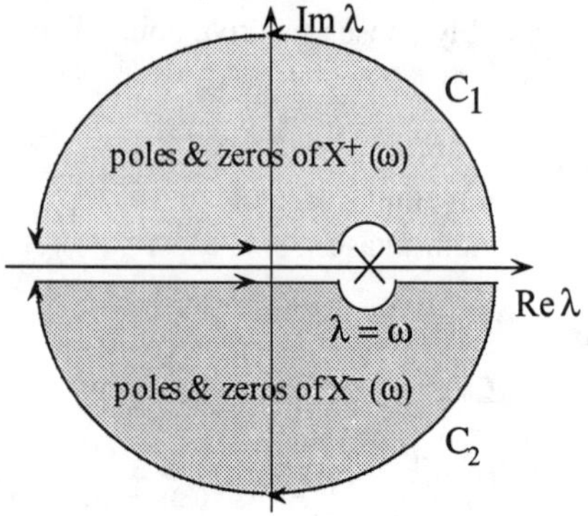

Hence,

$$\hat{X}(\omega) = X(\omega).[-j\ \mathrm{sgn}(\omega)] = [X^{+}(\omega) + X^{-}(\omega)][-j\ \mathrm{sgn}(\omega)]$$
$$= j\,[X^{-}(\omega) - X^{+}(\omega)]$$

where $X^{+}(\omega)$ is analytic in the LHP contour C_2, and $X^{-}(\omega)$ is analytic in the UHP contour C_1.

Referring to the contour, the following can be written:

$$\frac{1}{2\pi}\oint_{C_2} \frac{X^{+}(\lambda)}{-(\lambda-\omega)}\,d\lambda \tag{6}$$
$$= -\frac{1}{2}\left(-\text{ residue at pole } \lambda=\omega\right) = \frac{j\,X^{+}(\omega)}{2}$$

$$\frac{1}{2\pi}\oint_{C_1} \frac{X^{-}(\lambda)}{-(\lambda-\omega)}\,d\lambda \tag{7}$$
$$= \frac{1}{2}\left(-\text{ residue at pole } \lambda=\omega\right) = \frac{-j\,X^{-}(\omega)}{2}$$

Since the pole at $\lambda = \omega$ is shared by both contours C_1 and C_2, the contribution of the pole to each of the contours C_1 and C_2 can be deemed to be one half of the residue.

If x(t) is causal, then $x(t) = x^{+}(t)$ and if x(t) is anticausal, then $x(t) = x^{-}(t)$. Hence for causal x(t), $X(\omega) = X^{+}(\omega)$ and as a result,

$$\hat{X}(\omega) = X(\omega)*\frac{1}{\pi\,\omega} = \frac{1}{\pi}\int_{-\infty}^{\infty}\frac{X(\lambda)}{\omega-\lambda}\,d\lambda = j\,X(\omega) \tag{8}$$

As a consequence of the above result, for causal minimum phase functions with $X(\omega) = R(\omega) + j\,I(\omega)$,

$$R(\omega) = I(\omega) * \frac{1}{\pi\ \omega} : \ -I(\omega) = R(\omega) * \frac{1}{\pi\ \omega} \tag{9}$$

a classic relationship for causal minimum phase transfer functions showing that the real and imaginary parts of $X(\omega)$ cannot be specified independently.

Properties of x(t) and $\widehat{x}(t)$

1. $x(t)$ and $\widehat{x}(t)$ have the same energy if $x(t)$ is an energy signal and the same average power if $x(t)$ is a power signal.

2. $x(t)$ and $\widehat{x}(t)$ have the same autocorrelation function, i.e.,

$$\int_{-\infty}^{\infty} x(t)\, x(t+\tau)\, dt = \int_{-\infty}^{\infty} \hat{x}(t)\, \hat{x}(t+\tau)\, dt$$

$$= \int_{-\infty}^{\infty} x(t-\tau)\, x(t)\, dt = \int_{-\infty}^{\infty} \hat{x}(t-\tau)\, \hat{x}(t)\, dt \tag{10}$$

3. $x(t)$ and $\widehat{x}(t)$ are orthogonal, i.e.,

$$\int_{-\infty}^{\infty} x(t)\, \hat{x}(t)\, dt = 0 \tag{11}$$

4. Two HT operators in cascade behave as an ideal transformer, i.e., $x(t) \overset{\mathcal{H}}{\leftrightarrow} \hat{x}(t \overset{\mathcal{H}}{\leftrightarrow} -x(t)$

If $x(t)$ is a real causal time function, i.e., $x(t) = 0$, $t < 0$, with $X(\omega)$ satisfying the conditions,

1. Analyticity in the lower half λ-plane
2. $X(\lambda) = R(\lambda) + j\, I(\lambda)$: $R(\lambda)$ even; $I(\lambda)$ odd
3. $X(\lambda)$ analytic at $\lambda = 0$
4. limit of $X(\lambda)$ as $\lambda \to \infty$ is 0

then the Hilbert transform of $X(\omega)$ is given by:

$$X(\omega) \overset{\mathcal{H}}{\longleftrightarrow} \hat{\hat{X}}(\omega) = j\,X(\omega) = \frac{1}{\pi}\int_{-\infty}^{\infty} \frac{X(\lambda)}{\omega - \lambda}\, d\lambda \qquad (12)$$

In the case of time-shifted causal functions $x_c(t)$ defined by,

$$x_c(t) = x(t - c), \quad c > 0$$
$$X_c(\omega) = X(\omega)\, e^{-j\omega c} \text{ or, } X(\omega) = X_c(\omega)\, e^{j\omega c} \qquad (13)$$

The HT of $X_c(\omega)$ is given by:

$$X_c(\omega) \overset{\mathcal{H}}{\longleftrightarrow} \hat{\hat{X}}_c(\omega) = j\,X_c(\omega) = \frac{1}{\pi}\int_{-\infty}^{\infty} \frac{X_c(\lambda)}{\omega - \lambda}\, d\lambda \qquad (14)$$

Substituting $X(\omega) = X_c(\omega)\, e^{j\omega c}$ in eq.(12) we obtain,

$$j\,X_c(\omega)e^{j\omega c} = \frac{1}{\pi}\int_{-\infty}^{\infty} \frac{X_c(\lambda)\, e^{j\lambda c}}{\omega - \lambda}\, d\lambda\,, \quad \text{or,}$$

$$j\,X_c(\omega) = \frac{1}{\pi}\int_{-\infty}^{\infty} \frac{X_c(\lambda)\, e^{j(\lambda-\omega)c}}{\omega - \lambda}\, d\lambda \qquad (15)$$

A question now arises as to whether eqs.(14) and (15) are the same. We shall show that they are the same. Contour integration around C_1 in eq.(15) yields,

$$\frac{1}{\pi}\oint_{C_1} \frac{X_c(\lambda)e^{j(\lambda-\omega)c}}{\omega - \lambda}\, d\lambda = jX_c(\omega)$$

Conclusion: Given a measured value (at any point ω) for $X(\omega)$ or $X_c(\omega)$ of causal time functions $x(t)$ or $x_c(t)$, the HT of $X(\omega)$ or $X_c(\omega)$ can be expressed as a line integral by either eqs.(12) and (14), or eqs.(14) and (15).

SUMMARY-CONTINUOUS HILBERT TRANSFORMS

TIME DOMAIN:

Definitions: $x(t) \overset{\mathcal{H}}{\longleftrightarrow} \hat{x}(t$ $\qquad$ $x(t) \overset{\mathcal{F}}{\longleftrightarrow} X(\omega)$

Hilbert Transform: $\hat{x}(t) = x(t) * \dfrac{1}{\pi t} = \dfrac{1}{\pi}\displaystyle\int_{-\infty}^{\infty} \dfrac{x(\tau)}{t-\tau}\, d\tau$

Inverse Hilber Transform:

$$x(t) = -\hat{x}(t) * \frac{1}{\pi t} = -\frac{1}{\pi}\int_{-\infty}^{\infty} \frac{\hat{x}(\tau)}{t-\tau}\, d\tau$$

FT of $\hat{x}(t)$: $\qquad \hat{X}(\omega) = -j\, X(\omega)\, \mathrm{sgn}\, \omega$

FREQUENCY DOMAIN:

Definitions: $X(\omega) \overset{\mathcal{H}}{\longleftrightarrow} \hat{\hat{X}}(\omega) \overset{\mathcal{F}}{\longleftrightarrow} \hat{\hat{x}}(t)$

Hilbert Transform:

$$\hat{\hat{X}}(\omega) = X(\omega) * \frac{1}{\pi \omega} = \frac{1}{\pi}\int_{-\infty}^{\infty} \frac{X(\lambda)}{\omega-\lambda}\, d\lambda$$

Inverse Hilbert Transform

$$X(\omega) = -\hat{\hat{X}}(\omega) * \frac{1}{\pi \omega} = -\frac{1}{\pi}\int_{-\infty}^{\infty} \frac{\hat{\hat{X}}(\lambda)}{\omega-\lambda}\, d\lambda$$

Inverse Fourier Transform of $\hat{\hat{X}}(\omega)$

$$\hat{\hat{x}}(t) = \mathcal{F}^{-1}\left[X(\omega) * \frac{1}{j\omega}\right] = j\, x(t)\, \mathrm{sgn}\, t$$

Since $\hat{\hat{x}}(t) = j\,x(t)\,\mathrm{sgn}\,t$, for causal x(t), $x(t)\,\mathrm{sgn}\,t = x(t)$ and hence $\hat{\hat{x}}(t) = j\,x(t)$.

For causal x(t)

Hilbert Transform $\quad \hat{\hat{X}}(\omega) = j\,X(\omega) = \dfrac{1}{\pi}\displaystyle\int_{-\infty}^{\infty} \dfrac{X(\lambda)}{\omega - \lambda}\, d\lambda$

The flow chart of Hilbert Transform transformations is shown in the following figure.

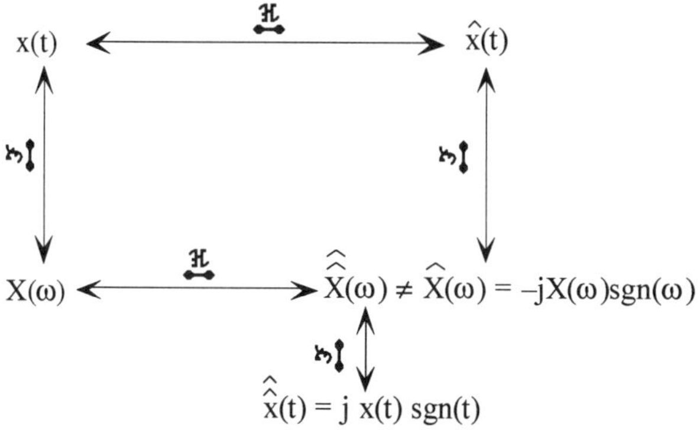

For causal x(t)

$\hat{\hat{X}}(\omega) = j\,X(\omega) = -\,\hat{X}(\omega)\mathrm{sgn}(\omega)$

Since $X(\lambda)$ is analytic in the lower half of the λ-plane for causal x(t), the contour C_2 in the lower half of the λ-plane is used in the evaluation of the causal x(t). Note that the contribution from the pole at $\lambda = \omega$ is only half the residue evaluated.

15. CONVERGENCE OF BILATERAL LAPLACE TRANSFORMS

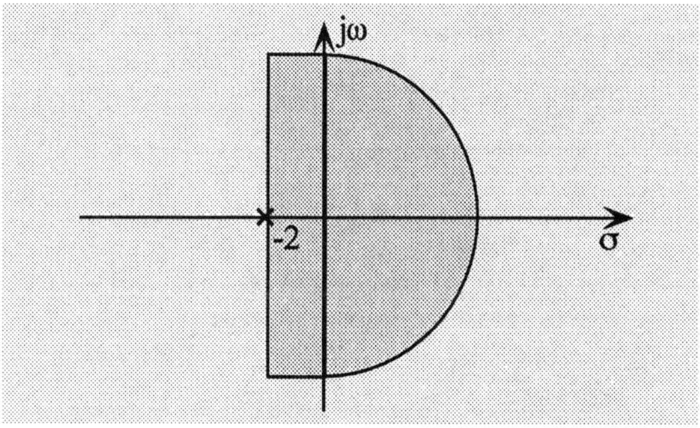

$$x(t) = e^{-2t}\,u(t)$$

$$X(s) = \frac{1}{s+2} : \text{ROC} : \text{Re}\{s\} > -2,\ X(j\omega) = \frac{1}{j\omega + 2}$$

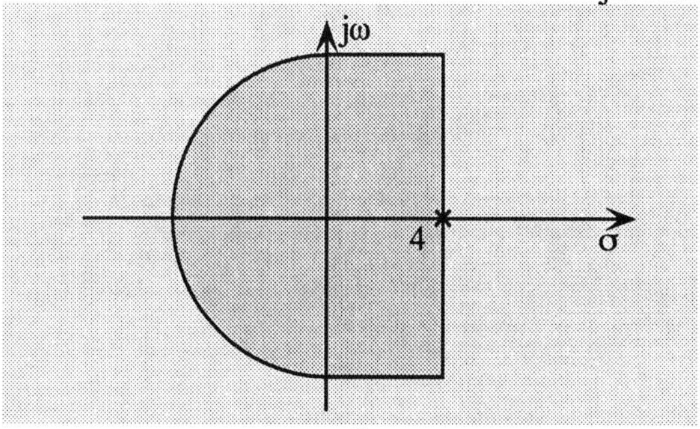

$$x(t) = e^{4t}\,u(-t)$$

$$X(s) = -\frac{1}{s-4} : \text{ROC} : \text{Re}\{s\} < 4,\ X(j\omega) = -\frac{1}{j\omega - 4}$$

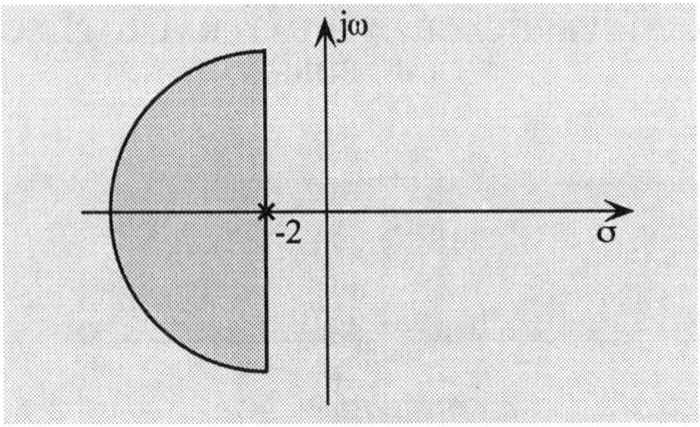

$$x(t) = e^{-2t}\,u(-t)$$

$$X(s) = -\frac{1}{s+2} : \text{ROC} : \text{Re}\{s\} < -2, \text{ No } X(j\omega)$$

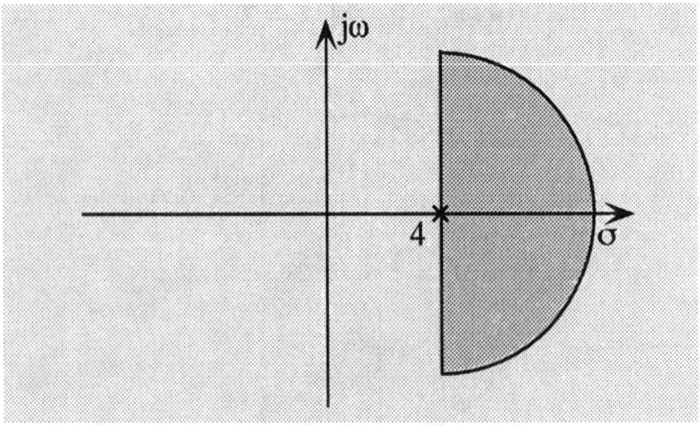

$$x(t) = e^{4t}\,u(t)$$

$$X(s) = \frac{1}{s-4} : \text{ROC} : \text{Re}\{s\} > 4, \text{ No } X(j\omega)$$

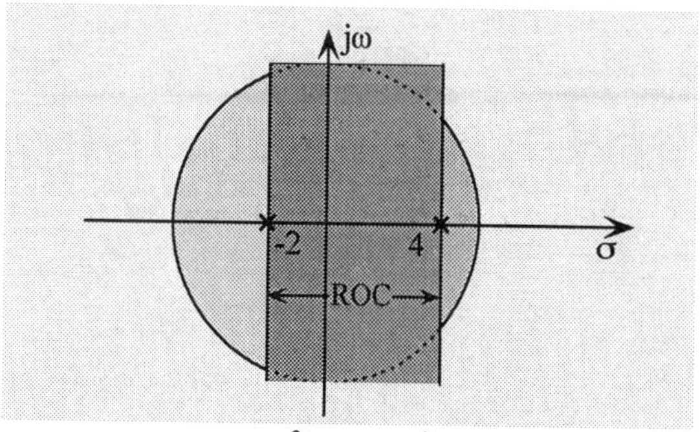

$$x(t) = e^{-2t}u(t) + e^{4t}u(-t)$$

$$X(s) = \frac{1}{s+2} - \frac{1}{s-4} : \text{ROC} : -2 < \text{Re}\{s\} < 4$$

$$X(j\omega) = \frac{1}{j\omega+2} - \frac{1}{j\omega-4}$$

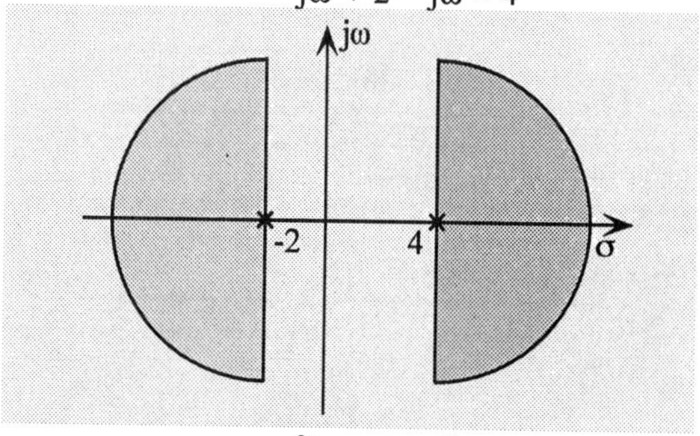

$$x(t) = e^{-2t}u(-t) + e^{4t}u(t)$$

$$\text{ROC} : \{\text{Re}\{s\} < -2\} \cap \{\text{Re}\{s\} > 4\} = \varnothing \text{ No LT, FT}$$

Further, the FT of x(t), X(jω) can be obtained from the LT X(s), only if the jω-axis is in the ROC of the s-plane. In the above examples FT exists only for examples 1,2,5.

16. PROPERTIES OF BILATERAL LAPLACE TRANSFORMS

DEFINITION: $\mathbf{x(t) \Leftrightarrow X(s) : y(t) \Leftrightarrow Y(s)}$

Bilateral Laplace Transform:

$$X(s) = \int_{-\infty}^{\infty} x(t)\, e^{-st} dt : \text{ROC} : R_x = \sigma_{x1} < \text{Re}\{s\} < \sigma_{x2}$$

$$Y(s) = \int_{-\infty}^{\infty} y(t)\, e^{-st} dt : \text{ROC} : R_y = \sigma_{y1} < \text{Re}\{s\} < \sigma_{y2}$$

ROC is always inside the innermost poles

Inverse Laplace Transform

$$x(t) = \frac{1}{2\pi j} \int_{\sigma - j\infty}^{\sigma + j\infty} X(s)\, e^{st} ds$$

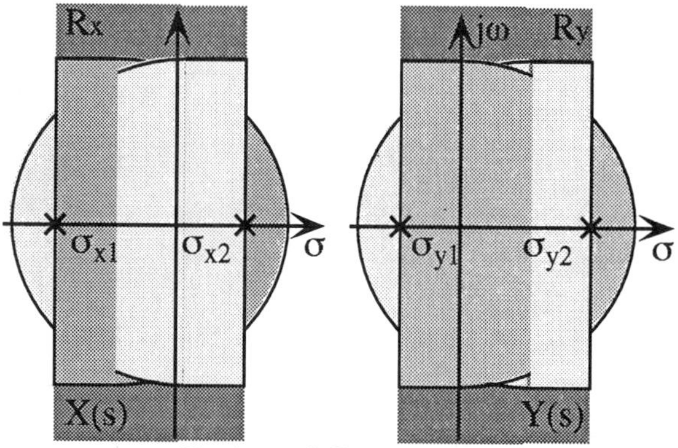

Regions of Convergence

Property		Signal	Transform	ROC
Properties				
1.	Linearity	$ax(t) + by(t)$	$aX(s) + bY(s)$	R_x « R_y
2.	Time Shift	$x(t - t_0)$	$e^{-st_0} X(s)$	R_x
3.	Frequency Shift	$x(t)\, e^{+s_0 t}$	$X(s - s_0)$	$R_x + \mathrm{Re}\{s_0\}$
4.	Scaling	$x(at)$	$\frac{1}{\mid a \mid} X\left(\frac{s}{a}\right)$	$\frac{R_x}{a}$
5.	Time Convolution	$x(t)*y(t)$	$X(s)Y(s)$	$R_x \cap R$
6.	Frequency Convolution	$x(t)y(t)$	$\frac{1}{2\pi j} X(s)*Y(s)$	$\{\sigma_{x1} + \sigma_{y1}, \sigma_{x2} + \sigma_{y2}\}$
7.	Time Differentiation	$\frac{d}{dt} x(t)$	$sX(s)$	R_x
8.	Frequency Differentiation	$-tx(t)$	$\frac{d}{ds} X(s)$	R_x
9.	Time Integration	$\int_{-\infty}^{\infty} x(t)dt$	$\frac{1}{s} X(s)$	$R_x \cap [\mathrm{Re}\{s\} > 0]$

17. UNILATERAL LAPLACE TRANSFORM (ULT) PAIRS

DEFINITION: $x(t) \overset{\mathcal{L}}{\longleftrightarrow} X(s)$

Unilateral Laplace transform: $X(s) = \int_{0-}^{\infty} x(t)\, e^{-st}\, dt$

Inverse Laplace transform: $x(t) = \frac{1}{2\pi j} \oint_C X(s)\, e^{st}\, ds$

Time Function-x(t)	Laplace Transform-X(s)
1. $u(t)$	$\frac{1}{s}$
2. $u(t) - u(t-a)$	$\frac{1 - e^{-as}}{s}$
3. $\delta(t)$	1
4. $\delta(t-a)$	e^{-as}
5. $t\, u(t)$	$\frac{1}{s^2}$
6. $t^n\, u(t)$	$\frac{n!}{s^{n+1}}$
7. $\left.\begin{matrix} 1, \cos \omega t \\ \sin\omega t \end{matrix}\right\} \quad -\infty < t < \infty$	No Laplace Transform
8. $e^{-\alpha t}\, u(t)$	$\frac{1}{s + \alpha}$

Time Function-x(t)	Laplace Transform- X(s)
9. $t\, e^{-\alpha t}\, u(t)$	$\dfrac{1}{(s+\alpha)^2}$
10. $t^n\, e^{-\alpha t}\, u(t)$	$\dfrac{n!}{(s+\alpha)^{n+1}}$
11. $\sin\omega t\, u(t)$	$\dfrac{\omega}{s^2+\omega^2}$
12. $\cos\omega t\, u(t)$	$\dfrac{s}{s^2+\omega^2}$
13. $\sin^2\omega t\, u(t)$	$\dfrac{2\omega^2}{s(s^2+4\omega^2)}$
14. $\cos^2\omega t\, u(t)$	$\dfrac{s^2+2\omega^2}{s(s^2+4\omega^2)}$
15. $e^{-\alpha t}\sin\omega t\, u(t)$	$\dfrac{\beta}{(s+\alpha)^2+\omega^2}$
16. $e^{-\alpha t}\cos\omega t\, u(t)$	$\dfrac{s+\alpha}{(s+\alpha)^2+\omega^2}$
17. $t\sin\omega t\, u(t)$	$\dfrac{2\omega s}{(s^2+\omega^2)^2}$
18. $t\cos\omega t\, u(t)$	$\dfrac{s^2-\omega^2}{(s^2+\omega^2)^2}$

18. COMPLEX CONVOLUTION
(Laplace Transform)

We shall demonstrate the important points in complex convolution or frequency convolution. Let x(t) and y(t) be Laplace transformable functions with Laplace transforms (LT) defined by,

$x(t) \overset{\mathcal{L}}{\longleftrightarrow} X(s)$, ROC: $\sigma_{x1} < \sigma_x$[1] $< \sigma_{x2}$

$y(t) \overset{\mathcal{L}}{\longleftrightarrow} Y(s)$, ROC: $\sigma_{y1} < \sigma_y$[1] $< \sigma_{y2}$ (1)

The LT of the product $z(t) = x(t)y(t)$ is desired using complex convolution and contour integration techniques. If the LT of z(t) is Z(s) with the region of convergence $\sigma_{z1} < \sigma_z < \sigma_{z2}$, then we have to determine σ_{z1} and σ_{z2} in terms of σ_{x1}, σ_{y1} and σ_{x2}, σ_{y2}. The existence of X(s) and Y(s) dictates the following relationships:

$$\int_{-\infty}^{\infty} \left| x(t)\, e^{-\sigma_x t} \right| dt < \infty \quad \text{for } \sigma_x > \sigma_{x1} \text{ and } \sigma_x < \sigma_{x2}$$

$$\int_{-\infty}^{\infty} \left| y(t)\, e^{-\sigma_y t} \right| dt < \infty \quad \text{for } \sigma_y > \sigma_{y1} \text{ and } \sigma_y < \sigma_{y2} \quad (2)$$

If $z(t) = x(t)y(t)$, then the existence of Z(s) is given by,

$$\int_{-\infty}^{\infty} \left| z(t) e^{-\sigma_z t} \right| dt = \int_{-\infty}^{\infty} \left| x(t)y(t) e^{-(\sigma_x + \sigma_y)t} \right| dt < \infty. \quad (3)$$

Hence, $\sigma_z = (\sigma_x + \sigma_y) > (\sigma_{x1} + \sigma_{y1})$ and $\sigma_z = (\sigma_x + \sigma_y) < (\sigma_{x2} + \sigma_{y2})$. Thus, the ROC for Z(s) is, $(\sigma_{x1} + \sigma_{y1}) < \sigma_z < (\sigma_{x_2} + \sigma_{y2})$ and, $\sigma_{z1} = (\sigma_{x1} + \sigma_{y1})$ and $\sigma_{z2} = (\sigma_{x2} + \sigma_{y2})$.

---

[1] Subscripts like σ_x, σ_y and σ_z have been added to $\sigma = \mathrm{Re}\{s\}$ to indicate the respective variables X, Y and Z.

We shall find an expression for Z(s) in terms of X(s) and Y(s) as a complex convolution.

$$\mathcal{L}\{x(t)y(t)\} = Z(s)$$

$$= \int_{-\infty}^{\infty} x(t)\left[\frac{1}{2\pi j}\int_{\sigma_p - j\infty}^{\sigma_p + j\infty} Y(p)\, e^{pt} dp\right] e^{-st} dt$$

$$= \frac{1}{2\pi j}\int_{\sigma_p - j\infty}^{\sigma_p + j\infty} Y(p)\left[\int_{-\infty}^{\infty} x(t)\, e^{-(s-p)t} dt\right] dp \qquad (4)$$

$$= \frac{1}{2\pi j}\int_{\sigma_p - j\infty}^{\sigma_p + j\infty} Y(p)\, X(s-p)\, dp = \frac{1}{2\pi j} Y(s)*X(s)$$

The integral on the right-hand side of eq.(4) has to be evaluated by contour integration around a suitably defined contour in the p-plane. Note that in these integrals, the limits of integration are from $(\sigma_p - j\infty)$ to $(\sigma_p + j\infty)$. This means that the *region of convergence* (ROC) in the *p-plane* where the line $p = \sigma_p$ has to be located is to be carefully determined. The ROC of Y(p) in the p-plane is $\sigma_{y1} < \sigma_p < \sigma_{y2}$. The corresponding ROC for X(s – p) in the p-plane is $\sigma_{x1} < (\sigma_z - \sigma_p) < \sigma_{x2}$. The union of these two ROCs results in $(\sigma_{x1} + \sigma_{y1}) < \sigma_z < (\sigma_{x2} + \sigma_{y2})$ in the s-plane, a result that was derived earlier.

In summary, to determine the ROC in the p-plane for the product Y(p)X(s–p) the following inequalities have to be satisfied.

1. $\sigma_{y1} < \sigma_p < \sigma_{y2}$
2. $\sigma_{x1} < (\sigma_z - \sigma_p) < \sigma_{x2}$, or
   $(\sigma_z - \sigma_{x2}) < \sigma_p < (\sigma_z - \sigma_{x1})$ with σ_z taking values as in (3) below
3. $(\sigma_{x1} + \sigma_{y1}) < \sigma_z < (\sigma_{x2} + \sigma_{y2})$

Inequalities (1) and (2) can be collapsed into a single inequality as given by the intersection of $\{\sigma_{y1}, \sigma_{y2}\} \cap \{\sigma_z - \sigma_{x2}, \sigma_z - \sigma_{x1}\}$ that results in,

4. $\max\{\sigma_{y1}, \sigma_z - \sigma_{x2}\} < \sigma_p < \min\{\sigma_{y2}, \sigma_z - \sigma_{x1}\}$ with σ_z ranging between $(\sigma_{x1} + \sigma_{y1})$ and $(\sigma_{x2} + \sigma_{y2})$ as in (3) above.

EXAMPLE

If $x(t) = 2te^{-t}u(-t) + e^{-4t}u(t)$ and $y(t) = e^{t}u(-t) + t\,u(t)$, find the LT of $z(t) = x(t)y(t)$ using complex convolution. Verify by actual multiplication.

The Laplace transforms X(s) and Y(s) with the corresponding regions of convergence are shown below:

$$X(s) = -\frac{2}{(s+1)^2} + \frac{1}{s+4} \qquad -4 < \sigma_x < -1 \text{ or } \begin{matrix}\sigma_{x1} = -4 \\ \sigma_{x2} = -1\end{matrix}$$

$$Y(s) = -\frac{2}{s-1} + \frac{1}{s^2} \qquad 0 < \sigma_y < 1 \text{ or } \begin{matrix}\sigma_{y1} = 0 \\ \sigma_{y2} = 1\end{matrix} \tag{5}$$

The regions of convergence for eq.(5) are shown in the following diagram.

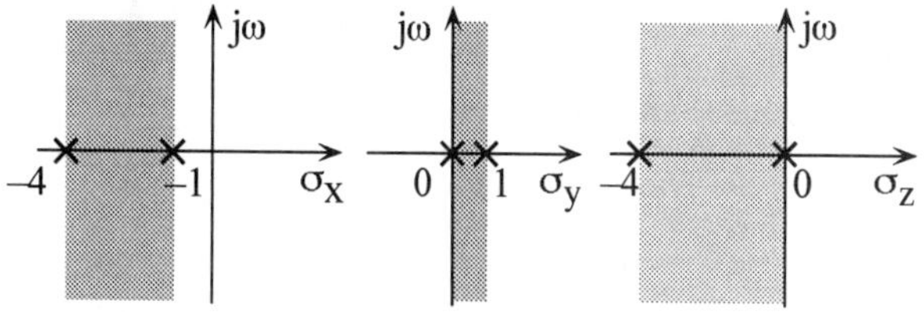

Regions of Convergence of X(s), Y(s), Z(s)

Corresponding to eq.(4) we have,

$\mathcal{L}\{x(t)y(t)\}$

$$= Z(s) = \frac{1}{2\pi j} X(s)*Y(s) = \frac{1}{2\pi j}\int_{\sigma_p - j\infty}^{\sigma_p + j\infty} Y(p)\, X(s-p)\, dp$$

with ROC given by: $-4 + 0 < \sigma_z < -1 + 1$ or, $-4 < \sigma_z < 0$. Substituting for Y(p) and X(s–p), Z(s) is written as:

$$Z(s) = \frac{1}{2\pi j}\int_{\sigma_p - j\infty}^{\sigma_p + j\infty} \underbrace{\left(-\frac{1}{p-1}+\frac{1}{p^2}\right)}\underbrace{\left(-\frac{2}{(s-p+1)^2}+\frac{1}{s-p+4}\right)} dp \quad (6)$$

We will now plot the poles of the function, $\left(-\frac{1}{p-1}+\frac{1}{p^2}\right)\left(-\frac{2}{(s-p+1)^2}+\frac{1}{s-p+4}\right)$ of eq.(6) in the complex p-plane as shown in the diagram below.

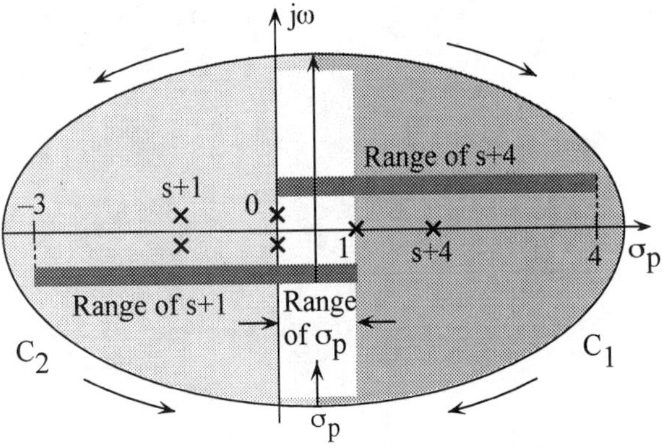

The ROC for the first term is $0 < \sigma_p < 1$, and the ROC for the second term is $(s+1) < \sigma_p < (s+4)$. The fixed poles are at $p = 0$ (double pole) and at $p = 1$. The moving poles are at $p = (s+1)^2$ (double pole) and at $p = s+4$ as shown in the diagram on the previous page.

Using the inequality, $\max\{\sigma_{y1}, (\sigma_z - \sigma_{x2})\} < \sigma_p < \min\{\sigma_{y2}, (\sigma_z - \sigma_{x1})\}$, we obtain three different cases depending upon the location of the moving poles, $p = (s+1)$ and $(s+4)$ corresponding to the ranges of $\mathrm{Re}\{s\} = \sigma_z$ between -4 and 0. These three cases are:

(i) $-3 < (s + 1) < -2$: $0 < (s + 4) < 1$
ROC: $0 < \sigma_p < (s + 4)$

(ii) $-2 < (s + 1) < 0$: $1 < (s + 4) < 3$
ROC: $0 < \sigma_p < 1$

(iii) $0 < (s + 1) < 1$: $3 < (s + 4) < 4$
ROC: $(s + 1) < \sigma_p < 1$

with $\mathrm{Re}\{s\} = \sigma_p$ satisfying $-4 < \mathrm{Re}\{s\} < 0$

In each of the three cases the right-hand contour C_1 contains the two poles at $p = 1$ and $p = (s+4)$ and the left-hand contour C_2 contains the two double poles at $p = 0$ and $p = (s+1)$. The ROC for $\sigma_p \in \{0,1\}$ for case (ii) is also shown in the diagram on the previous page. Having located σ_p, we can close the contour either on the left side or on the right side for evaluation of the integral in eq.(6). Since there is no time factor involved, evaluating the residues on either contour (C_1 or C_2) will yield exactly the same result (if the placement of σ_p is correct!). The residues at $p = 1$ and $p = s+4$ in contour C_1 are:

$$\mathrm{Res}\big|_{p=1} = \frac{2}{s^2} - \frac{1}{s+3}, \quad \mathrm{Res}\big|_{p=s+4} = \frac{1}{s+3} - \frac{1}{(s+4)^2} \tag{7}$$

Since the contour C_1 is in the clockwise direction we obtain Z(s) as the negative sum of residues given in eq.(7), that is,

$$Z(s) = -\frac{2}{s^2} + \frac{1}{(s+4)^2}, \quad -4 < \sigma_z < 0 \tag{8}$$

Evaluating the residues in the contour C_2 will also yield the same result. By direct multiplication we obtain

$$x(t)y(t) = 2t\, u(-t) + te^{-4t}\, u(t) \tag{9}$$

yielding the same LT as obtained in eq.(8).

19. PROPERTIES OF DISCRETE-TIME FOURIER SERIES (DTFS)

DEFINITION: **$x(n) \Leftrightarrow X(k)$**

Fundamental Frequency: $\Omega_0 = \frac{2\pi}{N}$

Synthesis Equation:

$$x(n) = \sum_{k=0}^{N-1} X(k)\, e^{j k\Omega_0 n}, \quad n = 0, \ldots, N-1$$

Analysis Equation:

$$X(k) = \frac{1}{N} \sum_{k=0}^{N-1} x(n)\, e^{-j k\Omega_0 n}, \quad k = 0, \ldots, N-1$$

Parseval's Relation:

$$\frac{1}{N} \sum_{n=0}^{N-1} |x(n)|^2 = \sum_{k=0}^{N-1} |X(k)|^2$$

Discrete Signal	Property	DTF Coefficient
$x(n) = x(n+mN)$	Periodic with period N	$X(k) = X(k+mN)$
$y(n) = y(n+mN)$	Periodic with period N	$Y(k) = Y(k+mN)$
$ax(n) + by(n)$	Linearity	$aX(k) + bY(k)$
$x(n-n_0)$	Time delay	$X(k)\, e^{-jk\Omega_0 n_0}$
$e^{jM\Omega_0 n}\, x(n)$	Frequency shifting	$X(k-M)$
$x^*(n)$	Complex conjugate	$X^*(N-k)$

Discrete Signal	Property	DTF Coefficient
$x(-n)$	Folding	$X(N-k)$
$x(n/m)$: Periodic with period mN	Scaling (only if n is a multiple of m)	$(1/m)X(k)$: Periodic with period mN
$\sum_{m=0}^{N-1} x(m)y(n-m)$	Time circular convolution	$NX(k)Y(k)$
$x(n)y(n)$	Frequency circular convolution	$\sum_{m=0}^{N-1} X(m)Y(k-m)$
$x(n) - x(n-1)$	Difference	$\{1 - e^{-k\Omega_0}\}\, X(k)$
$\sum_{m=-\infty}^{n} x(m) = \sum_{m=0}^{\infty} x(n-m)$	Summation (finite valued and periodic only if $X(0) = 0$)	$\dfrac{X(k)}{\left[1 - e^{-jk\Omega_0}\right]}$

SYMMETRY PROPERTIES:

$x(n)$	Real function	$X(k) = X^*(N-k)$ $Re\{X(k)\} = Re\{X(N-k)\}$ $Im\{X(k)\} = -Im\{X(N-k)\}$ $\lvert X(k)\rvert = \lvert X(N-k)\rvert$ $-X(k) = -X(N-k)$

Discrete Signal	Property	DTF Coefficient
$x_e(n) = Ev\{x(n)\}$	x(n) real	$Re\{X(k)\}$
$x_o(n) = Od\{x(n)\}$	x(n) real	$jIm\{X(k)\}$
$x(n) = x(-n)$	x(n) real even x(n) imaginary even	X(k) real even X(k) imaginary even
$x(n) = -x(-n)$	x(n) real odd x(n) imaginary odd	X(k) imaginary odd X(k) real odd
$x(n) = -x(n \pm N/2)$	Half-wave odd symmetry	Odd harmonics only
$x(n) = +x(n \pm N/2)$	Half-wave even symmetry (used only in a relative sense)	Even harmonics only

20. PROPERTIES OF DISCRETE-TIME FOURIER TRANSFORMS (DTFT)

DEFINITION: $x(n) \overset{\mathcal{F}}{\longleftrightarrow} X(\Omega)$

Repetition (Sampling) Frequency: $\Omega_s = 2\pi$

Synthesis Equation IDTFT:

$$x(n) = \frac{1}{2\pi}\int_0^{2\pi} X(\Omega)\, e^{j\Omega n}\, d\Omega$$

Analysis Equation DTFT:

$$X(\Omega) = \sum_{n=-\infty}^{\infty} x(n)\, e^{-j\Omega n}$$

Parseval's Relation:

$$\sum_{n=-\infty}^{\infty} |x(n)|^2 = \frac{1}{2\pi}\int_0^{2\pi} |X(\Omega)|^2 d\Omega$$

Discrete Signal	Property	DTFT
Finite Energy Signals		
$x(n)$	Periodic in frequency Ω with period 2π	$X(\Omega)=X(\Omega + 2\pi m)$
$y(n)$		$Y(\Omega)=Y(\Omega + 2\pi m)$
$ax(n) + by(n)$	Linearity	$aX(\Omega) + bY(\Omega)$
$x(n - n_0)$	Time shift	$e^{-j\Omega n_0}X(\Omega)$
$e^{j\Omega_0 n}x(n)$	Frequency shift	$X(\Omega - \Omega_0)$

Discrete Signal	Property	DTFT
$x^*(n)$	Complex conjugate	$X^*(-\Omega)$
$x(-n)$	Folding	$X(-\Omega)$
$x(n/m)$	Scaling (only if n is a multiple of m)	$X(m\Omega)$
$\sum_{m=-\infty}^{\infty} x(m)y(n-m)$	Time convolution	$X(\Omega)Y(\Omega)$
$x(n)y(n)$	Frequency convolution	$\frac{1}{2\pi}\int_0^{2\pi} X(\theta)Y(\Omega-\theta)d\theta$
$x(n) - x(n-1)$	Time Difference	$(1 - e^{-j\Omega})X(\Omega)$
$\sum_{m=-\infty}^{n} x(m) = \sum_{m=0}^{\infty} x(n-m)$	Summation	$\frac{1}{1-e^{-j\Omega}} X(\Omega) + \pi X(0) \sum_{k=-\infty}^{\infty} \delta(\Omega-2\pi k)$
$nx(n)$	Frequency differentiation	$j\frac{dX(\Omega)}{d\Omega}$
$x(n)$	Real function	$X(\Omega) = X^*(-\Omega)$ $\mathit{Re}\{X(\Omega)\} = \mathit{Re}\{X(-\Omega)\}$ $\mathit{Im}\{X(\Omega)\} = -\mathit{Im}\{X(-\Omega)\}$ $\lvert X(\Omega) = \lvert X(-\Omega)\rvert$ $-X(\Omega) = -X(-\Omega)$

Discrete Signal	Property	DTFT
Power and Discrete Singularity Signals		
$\delta(n)$	Discrete impulse (Kronecker delta)	$e^{j2\pi k}$
$1 = e^{j2\pi mn}$	Discrete unity	$2\pi \sum_{n=-\infty}^{\infty} \delta(\Omega - 2\pi m)$
$e^{j\Omega_0 n} = e^{j(\Omega_0 + 2\pi m)n}$	Frequency shifting	$2\pi \sum_{m=-\infty}^{\infty} \delta(\Omega - \Omega_0 - 2\pi m)$
$e^{jk\Omega_0 n} = e^{jk(\Omega_0 + 2\pi m)n}$	Frequency shifting	$2\pi \sum_{m=-\infty}^{\infty} \delta\left[\Omega - k(\Omega_0 - 2\pi m)\right]$
$\sum_{m=-\infty}^{\infty} \delta(n - mN)$	Discrete comb	$\frac{2\pi}{N} \sum_{m=-\infty}^{\infty} \delta\left(\Omega - m\frac{2\pi}{N}\right)$
$u(n)$	Discrete step	$\begin{cases} \frac{1}{1 - e^{-j\Omega}} + \\ \pi \sum_{k=-\infty}^{\infty} \delta(\Omega - 2\pi k) \end{cases}$
$x(n) = \sum_{k=0}^{N-1} X(k) e^{jk(2\pi n/N)}$	Periodic with period $2\pi/N$	$\sum_{m=-\infty}^{\infty} 2\pi X(k)\delta\left(\Omega - m\frac{2\pi}{N}\right)$

Discrete Signal	Property	DTFT
SYMMETRY PROPERTIES		
$x_e(n) = Ev\{x(n)\}$	x(n) real	$Re\{X(\Omega)$
$x_o(n) = Od(x(n)\}$	x(n) real	$jIm\{X(\Omega)\}$
$x(n) = x(-n)$	x(n) real even	$X(\Omega)$ real even
$x(n) = -x(-n)$	x(n) real odd	$X(\Omega)$ imaginary odd

21. PROPERTIES OF DISCRETE FOURIER TRANSFORMS (DFT)

DEFINITION: **$x(n) \overset{\mathcal{F}}{\longleftrightarrow} X(k)$**

Fundamental Frequency: $\Omega_0 = \frac{2\pi}{N}$: N roots of unity

$W_N = e^{j\Omega_0} : W_N^{-1} = e^{-j\Omega_0}$

Synthesis Equation (IDFT):

$$x(n) = \frac{1}{N}\sum_{k=0}^{N-1} X(k)\, W_N^{kn}, \; n = 0, \ldots, N-1$$

Analysis Equation (DFT):

$$X(k) = \sum_{n=0}^{N-1} x(n)\, W_N^{-kn}, \; k = 0, \ldots, N-1$$

Parseval's Relation:

$$\sum_{n=0}^{N-1} |x(n)|^2 = \frac{1}{N}\sum_{k=0}^{N-1} |X(k)|^2$$

Discrete Signal	**Property**	**DFT Coefficient**
$ax(n) + by(n)$	Linearity	$aX(k) + bY(k)$
$x(n - n_0)$	Time shifting	$e^{-jk\Omega_0 n_0}X(k)$
$e^{jM\Omega_0 n}x(n)$	Frequency shifting	$X(k - M)$
$nx^*(n)$	Complex conjugate	$X^*(-k)$
$x(-n)$	Folding	$X(-k)$
$x\left(\frac{n}{m}\right)$	Scaling (only if n is a multiple of m)	$X(mk)$

Discrete Signal	Property	DFT Coefficient
$\frac{1}{N}X(n)$	Duality	$x(-k)$
$\sum_{m=0}^{N-1} x(m)y(n-m)$	Time circular convolution	$X(k)Y(k)$
$x(n)y(n)$	Frequency circular convolution	$\frac{1}{N}\sum_{m=0}^{N-1} X(m)Y(k-m)$
$\sum_{n=0}^{N-1} x(n)$	Time summation	$X(0)$
$x(0)$	Frequency summation	$\frac{1}{N}\sum_{k=0}^{N-1} X(k)$

SYMMETRY PROPERTIES:

$x(n)$	Real function	$Re\{X(k) = Re\{X(-k)\}$ $Im\{X(k)\} = -Im\{X(-k)\}$ $X(k) = X^*(-k)$ $\|X(k)\|X(-k)\|$ $-X(k) = -X(-k)$
$x_e(n) = Ev\{x(n)\}$	x(n) real	$Re\{X(k)\}$
$x_o(n) = Od\{x(n)\}$	x(n) real	$jIm\{X(k)\}$
$x(n) = x(-n)$	x(n) real even x(n) imaginary even	X(k) real even X(k) imaginary even
$x(n) = -x(-n)$	x(n) real odd x(n) imaginary odd	X(k) imaginary odd X(k) real odd

Matrix Formulation of n-point IDFT

The analysis equation can be expressed in a matrix form connecting the time sequence {x(n)} to the frequency sequence {X(k)} as shown below:

$$\begin{bmatrix} x(0) \\ x(1) \\ x(2) \\ \vdots \\ x(N-1) \end{bmatrix} = \frac{1}{N} \begin{bmatrix} 1 & 1 & \cdots & 1 \\ 1 & W_N^{1} & \cdots & W_N^{(N-1)} \\ 1 & W_N^{2} & \cdots & W_N^{2(N-1)} \\ \vdots & \vdots & \vdots & \vdots \\ 1 & W_N^{(N-1)} & \cdots & W_N^{(N-1)^2} \end{bmatrix} \times \begin{bmatrix} X(0) \\ X(1) \\ X(2) \\ \vdots \\ X(N-1) \end{bmatrix}$$

Example - Matrix formulation of 8-point DFT

$$\begin{bmatrix} X(0) \\ X(1) \\ X(2) \\ \vdots \\ X(7) \end{bmatrix} = \begin{bmatrix} 1 & 1 & 1 & \cdots & 1 \\ 1 & \frac{1-j}{\sqrt{2}} & -j & \cdots & \frac{1+j}{\sqrt{2}} \\ 1 & -j & -1 & \cdots & j \\ \vdots & \vdots & \vdots & \vdots & \vdots \\ 1 & \frac{1+j}{\sqrt{2}} & j & \cdots & \frac{1-j}{\sqrt{2}} \end{bmatrix} \begin{bmatrix} x(0) \\ x(1) \\ x(2) \\ \vdots \\ x(7) \end{bmatrix}$$

22. GRAPHICAL DERIVATION OF DFT FROM CFT

The essential usefulness of the DFT is in its ability to approximate the continuous Fourier transform (CFT).

Discretization Intervals

Time domain: $$T = \frac{2\pi}{\omega_s}$$

Frequency domain: $$\Omega_0 = \frac{2\pi}{T_0}$$

N roots of unity: $$W_N = e^{j\frac{2\pi}{N}}$$

Truncation Intervals

Time domain: $$T_0 = NT$$

Frequency domain: $$\Omega_0 = N\Omega_0$$

N roots of unity: $$W_N^{-1} = e^{-j\frac{2\pi}{N}}$$

Synthesis equation:

$$x(nT) = \frac{1}{NT}\sum_{k=0}^{N-1} X(k\Omega_0)\, W_N^{kn} = \frac{\Omega_0}{2\pi}\sum_{k=0}^{N-1} X(k\Omega_0)\, W_N^{kn},\ 0 \le n < \frac{N}{2} \tag{1}$$

Analysis equation:

$$X(k\Omega_0) = T\sum_{n=0}^{N-1} x(nT)\, W^{-kn} = \frac{T_0}{N}\sum_{n=0}^{N-1} x(nT)\, W^{-kn},\ 0 \le n < \frac{N}{2} \tag{2}$$

Time Functions
A semi-infinite time function x(t) is given. It is necessary to discretize this function so that the discrete Fourier transform can be determined.
The sampling function $y_1(t)$ is a comb function with impulses of strength T and sampling interval T. T is chosen for acceptable aliasing error. $$y_1(t) = T \sum_{n=-\infty}^{\infty} \delta(t - nT) \quad (3)$$
$x_s(t) = x(t)y_1(t)$ has to be truncated in the time domain at a suitably determined time interval T_0. The sampled function $x_s(t)$ can be written as: $$x_s(t) = x(t)y_1(t) = T \sum_{n=-\infty}^{\infty} x(nT)\delta(t - nT) \quad (4)$$
The truncation filter w(t) is an ideal lp filter, $p_{T_0/2}(t)$, offset slightly by $-\varepsilon$ so that only (N – 1) samples are included. The truncation time T_0 is chosen to satisfy the frequency resolution $\Omega_0 = 2\pi/T_0$ with $T_0 = NT$. $$w(t) = p_{T_0/2}\left(t - \frac{T_0}{2} + \varepsilon\right) \quad (5)$$
The truncated sampled function $\tilde{x}_s(t)$ is given by: $$\tilde{x}_s(t) = x_s(t)w(t) = T \sum_{n=0}^{N-1} x(nT)\delta(t - nT) \quad (6)$$

Time Functions

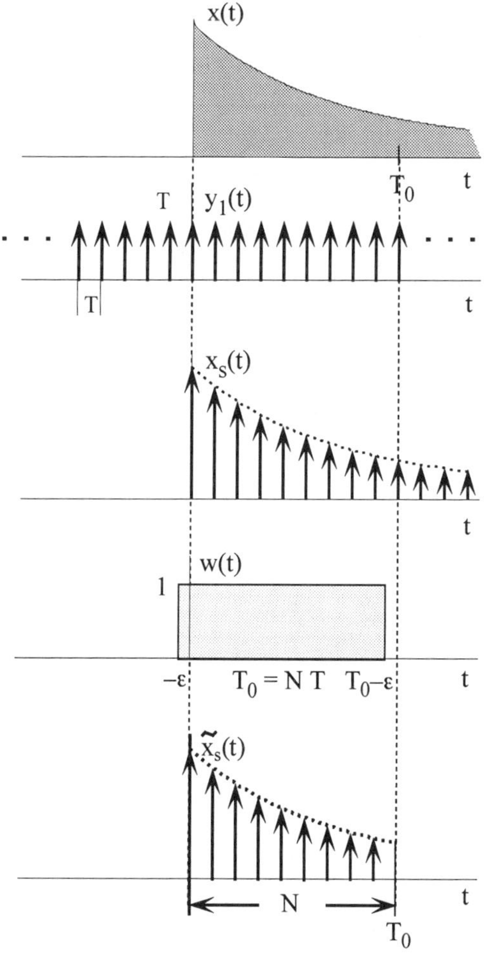

The corresponding time function $y_2(t)$ is again another comb function with spacing T_0.

$$y_2(t) = \sum_{n=-\infty}^{\infty} \delta(t - kT_0) : T_0 = \frac{2\pi}{\Omega_0} \tag{7}$$

The discrete time periodic sequence $\tilde{x}_{sp}(t)$ is:

$$\begin{aligned} \tilde{x}_{sp}(t) &= \tilde{x}_s(t)y_2(t) \\ &= T \sum_{k=-\infty}^{\infty} \left[\sum_{n=0}^{N-1} x(nT)\delta(t - nT - kT_0) \right] \end{aligned} \tag{8}$$

Note that $\Omega_0 T = \dfrac{2\pi}{N}$ so that $T = \dfrac{2\pi}{N\Omega_0}$

Time Functions

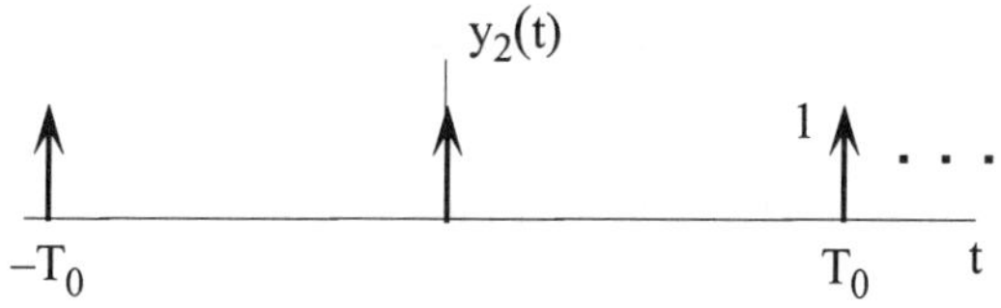

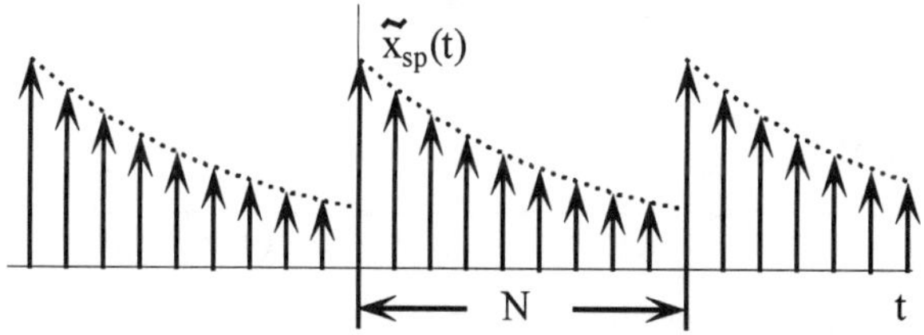

Frequency Functions
The corresponding continuous time Fourier transform, $X(\omega)$ is also known.
The Fourier transform $Y_1\omega)$ is also a comb function given by: $$Y_1(\omega) = 2\pi \sum_{n=-\infty}^{\infty} \delta(\omega - n\omega_s) : \omega_S = \frac{2\pi}{T} \qquad (9)$$
The Fourier transform $X_s(\omega)$ is a continuous periodic function with period $\omega_s = \frac{2\pi}{T}$. T is so chosen that the aliasing error is within acceptable limits. $$X_s(\omega) = \frac{1}{2\pi} X(\omega)*Y_1(\omega) = \sum_{n=-\infty}^{\infty} X(\omega - n\omega_s) \qquad (10)$$
The Fourier transform $W(\omega)$ is a Sa(.) function with zero crossings at the desired frequency resolution $\Omega_0 = \frac{2\pi}{T_0}$. $$W(\omega) = T_0\, e^{-j\omega(T_0/2 - \varepsilon)}\, Sa\left(\frac{\omega T_0}{2}\right) \qquad (11)$$
The corresponding FT, $\widetilde{X}_s(\omega)$ is obtained by directly taking the FT of $\widetilde{x}_s(t)$. $\widetilde{X}_s(\omega)$ is still a continuous periodic function that has to be discretized by multiplying with a comb function $Y_2(\omega)$. $$\widetilde{X}_s(\omega) = X_s(\omega)*W(\omega) = T \sum_{n=0}^{N-1} x(nT)\, e^{-j\omega nT} \qquad (12)$$

Frequency Functions

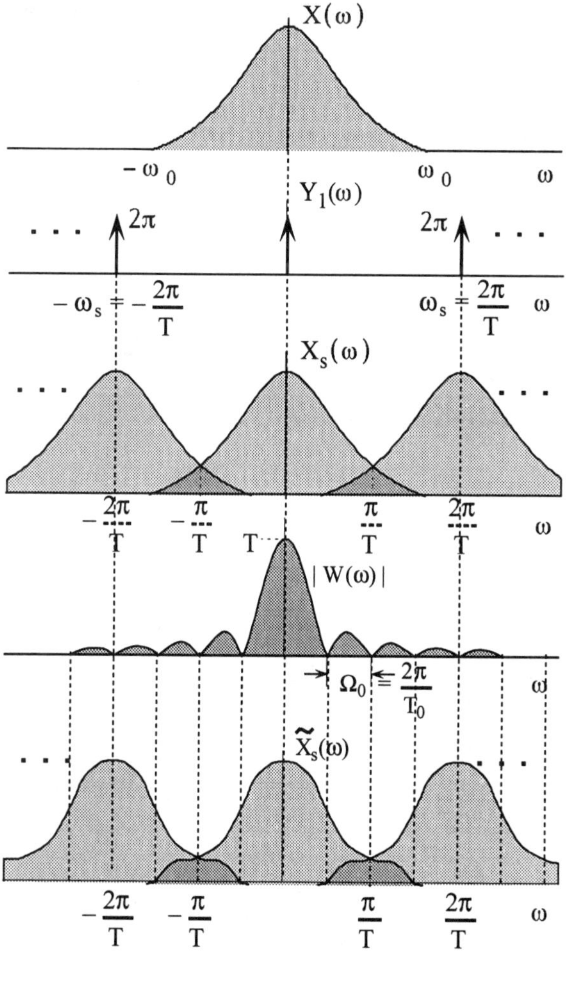

Frequency Functions

The sampling function $Y_2(\omega)$ is a comb function with sampling interval Ω_0 set at the desired frequency resolution and impulse strength Ω_0.

$$Y_2(\omega) = \Omega_0 \sum_{n=-\infty}^{\infty} \delta(\omega - k\Omega_0) \qquad (13)$$

The discrete frequency periodic sequence $\widetilde{X}_{sp}(\omega)$ is

$$\begin{aligned}\widetilde{X}_{sp}(\omega) &= \widetilde{X}_s(\omega) * Y_2(\omega) \\ &= \Omega_0 \sum_{n=0}^{N-1} X_s(k\Omega_0)\, \delta(\omega - k\Omega_0)\end{aligned} \qquad (14)$$

$$\Omega_0 = \frac{2\pi}{NT}$$

Frequency Functions

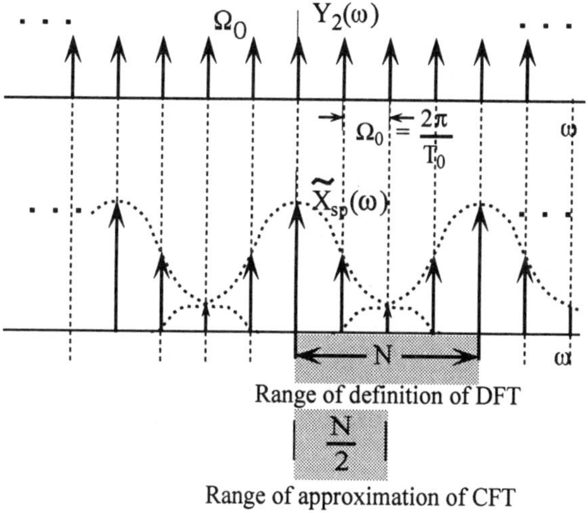

We have successfully approximated the continuous time signal x(t) and its Fourier transform X(ω) by a discrete time sequence $\tilde{x}_{sp}(t)$ and a discrete frequency sequence $\tilde{X}_{sp}(\omega)$. Since,

$$\tilde{X}_{sp}(\omega) = \tilde{X}_s(\omega)Y_2(\omega)$$
$$= \left[T\sum_{n=0}^{N-1} x(nT)e^{-j\omega nT}\right]\left[\Omega_0 \sum_{k=-\infty}^{\infty} \delta(\omega - k\Omega_0)\right] \quad (15)$$

we can rewrite this equation as

$$\tilde{X}_{sp}(\omega)$$
$$= \underbrace{\frac{1}{N}\left[\sum_{n=0}^{N-1} x(nT)e^{-jk\Omega_0 nT}\right]}\left[2\pi \sum_{k=-\infty}^{\infty} \delta(\omega - k\Omega_0)\right] \quad (16)$$

and interpret the quantity

$$\frac{1}{N}\left[\sum_{n=0}^{N-1} x(nT)e^{-jk\Omega_0 nT}\right]$$

as the Fourier coefficient $\tilde{X}_{spk}$ of the discrete time periodic function $\tilde{x}_{sp}(t)$, which is shown in the figure on page 124. But $\tilde{x}_{sp}(t)$ is the discrete approximation of x(t) in the interval $\{0 \le t < T_0\}$. Thus, we have the following discrete-time Fourier series given by:

$$\left[\begin{aligned} \tilde{X}_{spk} &\cong \frac{1}{N}\sum_{n=0}^{N-1} x(nT)e^{-jk\Omega_0 nT} \\ x(nT) &\cong \sum_{k=0}^{N-1} \tilde{X}_{spk}e^{jk\Omega_0 nT} \end{aligned}\right] \text{in the interval } 0 \le t < T_0 \quad (17)$$

We have to relate these results to the Fourier transform, X(ω) of the continuous time function x(t). Using the Fourier series ⇒ Fourier transform relationships we can write:

$$\lim_{T_0 \to \infty} \left[T_0 \, \widetilde{X}_{spk} \right] = \widetilde{X}_s(\omega) \cong X(\omega), \; -\frac{\omega_s}{2} < \omega < \frac{\omega_s}{2} \tag{18}$$

or conversely,

$$\frac{X(k\Omega_0)}{T_0} \cong \widetilde{X}_{spk}, \; -\frac{\omega_s}{2} < \omega < \frac{\omega_s}{2} \tag{19}$$

Substituting this result in the DTFS expressions Eq.(15) we obtain the desired DFT approximation to the CFT as:

$$\left. \begin{aligned} X(k\Omega_0) &\cong \frac{T_0}{N} \sum_{n=0}^{N-1} x(nT)\, e^{-jk\Omega_0 nT} \\ x(nT) &\cong \frac{1}{T_0} \sum_{k=0}^{N-1} X(k\Omega_0)\, e^{jk\Omega_0 nT} \end{aligned} \right\} \text{for} \begin{cases} 0 \le n < \dfrac{N}{2} \\ 0 \le k < \dfrac{N}{2} \end{cases} \tag{20}$$

and substituting $T_0 = NT$ in eq.(18) we can also obtain the DFT pair as:

$$\left. \begin{aligned} X(k\Omega_0) &\cong T \sum_{n=0}^{N-1} x(nT)\, e^{-jk\Omega_0 nT} \\ x(nT) &\cong \frac{1}{NT} \sum_{k=0}^{N-1} X(k\Omega_0)\, e^{jk\Omega_0 nT} \end{aligned} \right\} \text{for} \begin{cases} 0 \le n < \dfrac{N}{2} \\ 0 \le k < \dfrac{N}{2} \end{cases} \tag{21}$$

Equations (20, 21) correspond to eqs.(1, 2).

Note that if $T = 1$ in eqs.(20, 21), we obtain expressions for DTFS similar to those in section 19, except that the term $\frac{1}{N}$ appears in the $x(n)$ term rather than in $X(k)$.

23. ANALYTICAL DERIVATION OF FFT ALGORITHM

General Method

The definition of DFT from Section 21 is

$$X(k) = \sum_{n=0}^{N-1} x(n)\, W_N^{-kn} \quad k = 0, \cdots, N-1 \tag{1}$$

Assume that $N = 2^m$ where m is an integer. X(k) can be split into odd and even sequences as given by:

$$\begin{aligned} X(k) = & \sum_{n=0}^{\frac{N}{2}-1} x(2n)\, W_N^{-k2n} \\ & + \sum_{n=0}^{\frac{N}{2}-1} x(2n+1)\, W_N^{-k(2n+1)} \quad k = 0, \cdots, N-1 \end{aligned} \tag{2}$$

Eq. (2) can be compactly expressed as

$$X(k) = \sum_{n_0=0}^{1} \sum_{n=0}^{\frac{N}{2}-1} x(2n+n_0)\, W_N^{-k(2n+n_0)} \tag{3}$$

$$k = 0, \cdots, N-1$$

Eq.(3) is basically eq.(1) with n replaced by $2n + n_0$ with a double summation replacing the single summation. Eq.(3) can again be split into odd and even sequences, and the same pattern can be repeated by replacing n by $2n+n_1$ and adding another summation as given by:

$$X(k) = \sum_{n_0=0}^{1} \sum_{n_1=0}^{1} \sum_{n=0}^{\frac{N}{4}-1} x\,(2^2 n+2n_1+n_0)\, W_N^{-k(2^2 n+2n_1+n_0)}$$

$$k = 0, 1, \ldots, N-1 \tag{4}$$

We repeat these successive divisions into odd and even sequences m times so that all the $N = 2^m$ points are exhausted. This results in:

$$X(k) = \sum_{n_0=0}^{1} \sum_{n_1=0}^{1} \cdots \sum_{n_{m-1}=0}^{1} X(2^{m-1} n_{m-1} + \ldots + 2n_1 + n_0)$$

$$\times\ W_N^{-k(2^{m-1} n + \cdots + 2n_1 + n_0)} \qquad (5)$$

$$k = 0, \cdots, N-1$$

We can now represent in eq.(5) both n and k in binary notation as

$$n = n_{m-1}2^{m-1} + n_{m-2}2^{m-2} + \ldots + n_1 2^1 + n_0 2^0$$
$$k = k_{m-1}2^{m-1} + k_{m-2}2^{m-2} + \ldots + k_1 2^1 + k_0 2^0 \qquad (6)$$

Note that the successive division into odd and even sequences has automatically converted n into a binary representation. Substituting for k from eq.(6) into eq.(5) we obtain a complete binary representation,

$$X\left(k_{m-1}, k_{m-2}, \cdots, k_1, k_0\right)$$

$$= \sum_{n_0=0}^{1} \cdots \sum_{n_{m-1}=0}^{1} \Big[x\left(n_{m-1}, \cdots, n_1 + n_0\right) \qquad (7)$$

$$\times\ W_N^{-(2^{m-1} k_{m-1} + \cdots + k_0)(2^{m-1} n_{m-1} + \cdots + n_0)} \Big]$$

Depending upon how the $W_N^{-(.)(.)}$ term is simplified, we have two different types of algorithms: *decimation in time* if the binary n is factored and *decimation in frequency* if the binary k is factored.

Decimation in Time

The $W_N^{-(.)(.)}$ term is factored in terms of the binary n as follows:

$$W_N^{-(2^{m-1}k_{m-1}+\cdots+2k_1+k_0)(2^{m-1}n_{m-1}+\cdots+2n_1+n_0)}$$

$$= \left[W_N^{-(2^{m-1}k_{m-1}+\cdots+2k_1)2^{m-1}n_{m-1}}\right]$$

$$\times W_N^{-k_0 2^{m-1}n_{m-1}}$$

$$\times \left[W_N^{-(2^{m-1}k_{m-1}+\cdots)2^{m-2}n_{m-2}}\right]$$

$$\times W_N^{-(2k_1+k_0)2^{m-2}n_{m-2}}$$

$$\vdots$$

$$\times \left[W_N^{-(2^{m-1}k_{m-1})2^{1}n_1}\right] \tag{8}$$

$$\times W_N^{-(2^{m-2}k_{m-2}+\cdots+2k_1+k_0)2^{1}n_1}$$

$$\times W_N^{-(2^{m-1}k_{m-1}+\cdots+2k_1+k_0)2^{0}n_0}$$

The expressions within the square braces in the RHS of eq.(8) are always 1 since $W^{N\times \text{any integer}}$ is 1. Hence, eq.(8) can be rewritten as:

$$W_N^{-(2^{m-1}k_{m-1}+\cdots+2k_1+k_0)(2^{m-1}n_{m-1}+\cdots+2n_1+n_0)}$$

$$= W_N^{-k_0 2^{m-1}n_{m-1}}.W_N^{-(2k_1+k_0)2^{m-2}n_{m-2}}\;\cdots\cdots$$

$$\times W_N^{-(2^{m-2}k_{m-2}+\cdots+2k_1+k_0)2^{1}n_1} \tag{9}$$

$$\times W_N^{-(2^{m-1}k_{m-1}+\cdots+2k_1+k_0)2^{0}n_0}$$

Substituting eq.(9) into eq.(7) we obtain:

$$X(k_{m-1}, k_{m-2}, \cdots, k_1, k_0) =$$

$$\sum_{n_0=0}^{1} \cdots \sum_{n_{m-1}=0}^{1} x\left(n_{m-1}, n_{m-2}, \cdots, n_1, n_0\right)$$

$$\times W_N^{-k_0 2^{m-1} n}$$

$$\times W_N^{-(2k_1+k_0)2^{m-2} n_{m-2}} \times \ldots \tag{10}$$

$$\times W_N^{-(2^{m-1} k_{m-1}+\cdots+2k_1+k_0)2^0 n_0}$$

We can now define the following m iterations:

$$x_1(k_0, n_{m-2}, \cdots, n_1, n_0)$$

$$= \sum_{n_{m-1}=0}^{1} x(n_{m-1}, n_{m-2}, \cdots, n_1, n_0)$$

$$\times W_N^{-k_0 2^{m-1} n_{m-1}}$$

$$x_2(k_0, k_1, \cdots, n_1, n_0)$$

$$= \sum_{n_{m-2}=0}^{1} x_1(k_0, n_{m-2}, \cdots, n_1, n_0)$$

$$\times W_N^{-(2k_1+k_0)2^{m-2} n_{m-2}}$$

$$\vdots$$

$$x_{m-1}(k_0, k_1, \cdots, k_{m-2}, n_0)$$

$$= \sum_{n_1=0}^{1} x_{m-2}(k_0, k_1, \cdots, n_1, n_0)$$

$$\times W_N^{-(2^{m-2} k_{m-2}+\cdots+2k_1+k_0)2^1 n_1}$$

$$x_m(k_0,k_1,\cdots,k_{m-2},k_{m-1})$$

$$= \sum_{n_0=0}^{1} x_{m-1}(k_0,k_1,\cdots,k_{m-2},n_0)$$

$$\times W_N^{-(2^{m-1}k_{m-1}+\cdots+2k_1+k_0)\,2^0 n_0} \qquad (11)$$

$$X(k_{m-1},k_{m-2},\cdots,k_1,k_0)$$

$$= x_m\,(k_0,k_1,\cdots,k_{m-2},k_{m-1})$$

Decimation in Frequency

The $W_N^{-(.)(.)}$ term is now expanded in terms of the binary k to obtain the decimation in frequency algorithm.

$$W_N^{-(2^{m-1}k_{m-1}+\cdots+2k_1+k_0)(2^{m-1}n_{m-1}+\cdots+2n_1+n_0)}$$

$$= \left[W_N^{-(2^{m-1}k_{m-1})}\right] \times W_N^{-(2^{m-1}n_{m-1}+\cdots+2n_1+n_0)}$$

$$\times \left[W_N^{-(2^{m-2}k_{m-2})}\right] \times W_N^{-(2^{m-1}n_{m-1}+\cdots+2n_1+n_0)}$$

$$\vdots$$

$$\times \left[W_N^{-(2^1k_1)}\right] \times W_N^{-(2^{m-1}n_{m-1}+\cdots+2n_1+n_0)}$$

$$\times \left[W_N^{-(2^0k_0)}\right] \times W_N^{-(2^{m-1}n_{m-1}+\cdots+2n_1+n_0)} \qquad (12)$$

Applying $2^m = N$, and using the periodicity condition $W_N^{(kn+mN)} = W_N^{kn}$, eq.(12) can be simplified to:

$$W_N^{-(2^{m-1}k_{m-1}+\cdots+2k_1+k_0)(2^{m-1}n_{m-1}+\cdots+2n_1+n_0)}$$

$$= W_N^{-n_0 2^{m-1}k_{m-1}} . W_N^{-(2n_1+n_0)\,2^{m-2}k_{m-2}} \times \cdots$$

$$\times W_N^{-(2^{m-2}n_{m-2}+\cdots+2n_1+n_0)\,2^1k_1}$$

$$\times W_N^{-(2^{m-1}n_{m-1}+\cdots+2n_1+n_0)\,2^0k_0} \tag{13}$$

Substituting eq.(13) into eq.(7) we obtain after simplifications:

$$X(k_{m-1}, k_{m-2}, \cdots, k_1, k_0) =$$

$$\sum_{n_0=0}^{1} \cdots \sum_{n_{m-1}=0}^{1} x(n_{m-1}, n_{m-2}, \cdots, n_1, n_0)$$

$$\times W_N^{-(2^{m-1}n_{m-1}+\cdots+2n_1+n_0)\,2^0k_0} \tag{14}$$

$$\times W_N^{-(2^{m-2}n_{m-2}+\cdots+2n_1+n_0)2^1k_1} \cdots$$

$$\times W_N^{-(2n_1+n_0)2^{m-2}k_{m-2}} . W_N^{-n_0\,2^{m-1}k_{m-1}}$$

Once again, we can define the following iterations:

$$x_1(k_0, n_{m-2}, \cdots, n_1, n_0)$$

$$= \sum_{n_{m-1}=0}^{1} x(n_{m-1}, n_{m-2}, \cdots, n_1, n_0)$$

$$\times W_N^{-(2^{m-1}n_{m-1}+\cdots+2n_1+n_0)2^0k_0}$$

$$x_2(k_0, k_1, \cdots, n_1, n_0)$$

$$= \sum_{n_{m-2}=0}^{1} x_1(k_0, n_{m-2}, \cdots, n_1, n_0)$$

$$\times W_N^{-(2^{m-2}n_{m-2}+\cdots+2n_1+n_0)2^1k_1}$$

$$\vdots$$

$$x_{m-1}(k_0,k_1,\cdots,k_{m-2},n_0) = \sum_{n_1=0}^{1} x_{m-2}(k_0,k_1,\cdots,n_1,n_0) \times W_N^{-(2n_1+n_0)2^{m-2}k_{m-2}}$$

$$x_m(k_0,k_1,\cdots,k_{m-2},k) = \sum_{n_0=0}^{1} x_{m-1}(k_0,k_1,\cdots,k_{m-2},n_0) \times W_N^{-n_0 2^{m-1}k_{m-1}} \tag{15}$$

$$X(k_{m-1},k_{m-2},\cdots,k_1,k_0) = x_m(k_0,k_1,\cdots,k_{m-2},k_{m-1})$$

We can see clearly in both cases that when the input sequence is in the natural order $x(n_{m-1}, n_{m-2}, \ldots n_1, n_0)$, the output DFT sequence is in the bit-reversed order $X(k_0, k_1, \ldots, k_{m-2}, k_{m-1})$. Thus the output sequence has to be unscrambled using a bit-reversal algorithm.

Example

Decimation in time

As an example, we shall take $N = 2^3 = 8$. This gives m = 3, giving 3 iterations for eq.(11). Equations similar to eqs.(10, 11) can be written for N = 8.

$$X(k_2, k_1, k_0) = \sum_{n_0=0}^{1} \sum_{n_1=0}^{1} \sum_{n_2=0}^{1} x(n_2, n_1, n_0) W_N^{-k_0 2^2 n_2} \times W_N^{-(2k_1 + k_0)\, 2^1 n_1} \times W_N^{-(2^2 k_2 + 2^1 k_1 + k_0)\, 2^0 n_0} \tag{16}$$

and the three iterations are:

$$\begin{aligned} x_1(k_0,n_1,n_0) &= \sum_{n_2=0}^{1} x(n_2,n_1,n_0) W_N^{-k_0 2^2 n_2} \\ x_2(k_0,k_1,n_0) &= \sum_{n_1=0}^{1} x_1(k_0,n_1,n_0) \times W_N^{-(2k_1+k_0)2^1 n_1} \\ x_3(k_0,k_1,k_2) &= \sum_{n_0=0}^{1} x_2(k_0,k_1,n_0) \times W_N^{-(2^2 k_2+2k_1+k_0)2^0 n_0} \end{aligned} \tag{17}$$

$$X(k_2,k_1,k_0) = x_3(k_0,k_1,k_2)$$

Decimation in frequency

Similarly, for decimation in frequency we have,

$$X(k_2,k_1,k_0) = \sum_{n_0=0}^{1} \sum_{n_1=0}^{1} \sum_{n_2=0}^{1} x(n_2,n_1,n_0) W_N^{-n_0 2^2 k_2} \times W_N^{-(2n_1+n_0)\, 2^1 k_1} \times W_N^{-(2^2 n_2+2^1 n_1+n_0)\, 2^0 k_0} \tag{18}$$

and the corresponding three iterations are,

$$x_1(k_0,n_1,n_0) = \sum_{n_2=0}^{1} x(n_2,n_1,n_0) W_N^{-(2^2 n_2+2n_1+n_0) 2^0 k_0}$$

$$x_2(k_0,k_1,n_0) = \sum_{n_1=0}^{1} x_1(k_0,n_1,n_0) W_N^{-(2n_1+n_0) 2^1 k_1} \tag{19}$$

$$x_3(k_0,k_1,k_2) = \sum_{n_0=0}^{1} x_2(k_0,k_1,n_0) W_N^{-n_0 2^2 k_2}$$

$$X(k_2,k_1,k_0) = x_3(k_0,k_1,k_2)$$

The bit reversals are evident in both cases.

The signal flow diagram using decimation in time for the 8-point FFT of the above example is shown in the diagram below.

8-pt FFT-Decimation in Time

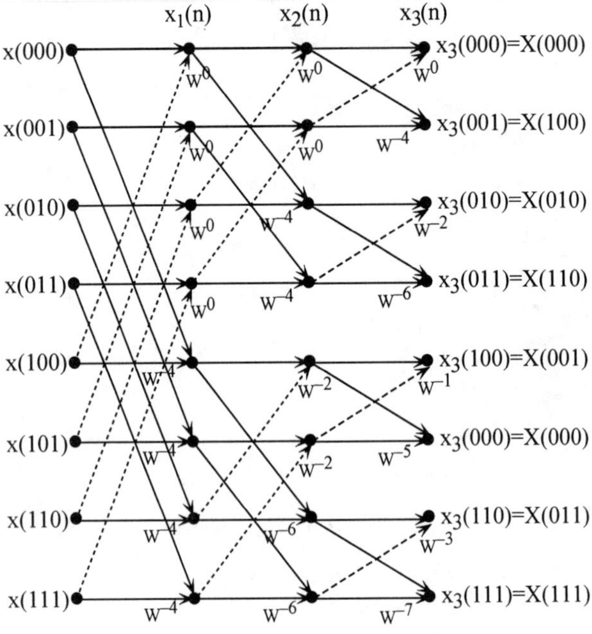

Similarly, the signal flow diagram using decimation in frequency for the 8-point FFT of the above example is shown in the diagram below.

8-pt FFT-Decimation in Frequency

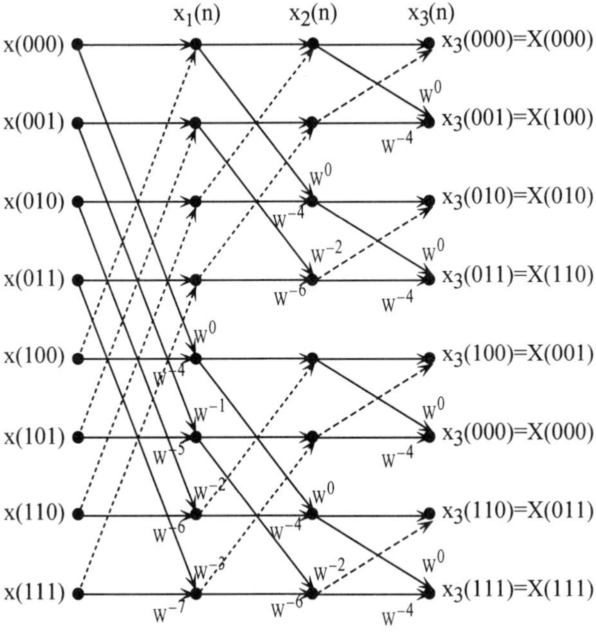

24. CONVERGENCE OF BILATERAL Z-TRANSFORMS

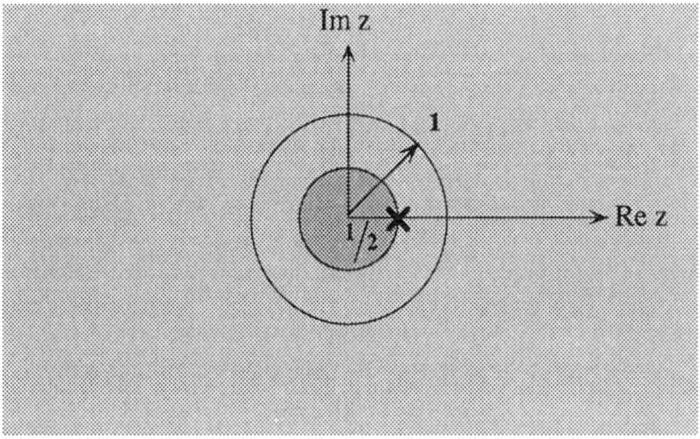

$$x(n) = \left(\frac{1}{2}\right)^n u(n)\,,\; X(z) = \frac{z}{z - \frac{1}{2}}\,,\; \text{ROC: } |z| > \frac{1}{2}$$

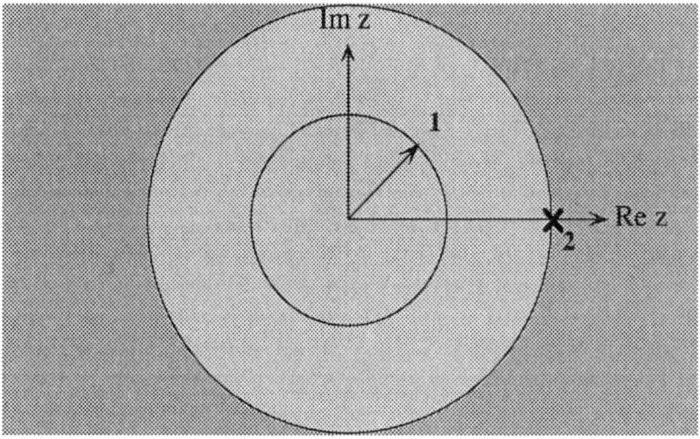

$$x(n) = -2^n u(-n-1)\,,\; X(z) = \frac{z}{z-2}\,,\; \text{ROC : } |z| < 2$$

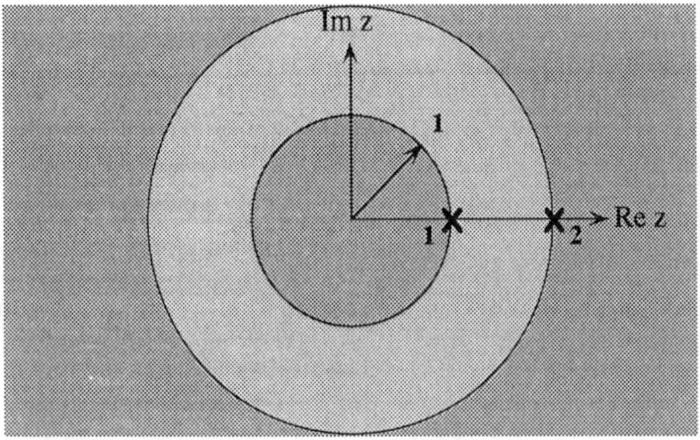

$$x(n) = u(n) - 2^n u(-n-1) \, , \; X(z) = \frac{z}{z-1} + \frac{z}{z-2}$$

ROC: $1 < |z| < 2$

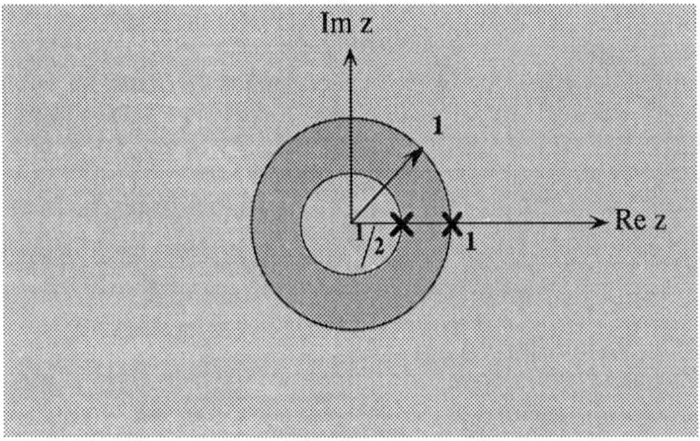

$$x(n) = u(n) + 2^{-n} u(-n-1)$$

ROC: $\left[|z| < \frac{1}{2}\right] \cap \left[|z| > 1\right] = \varnothing \Rightarrow$ No ZT

Laplace Transform (LT) to z-Transform (ZT) (impulse invariance method)

A mapping from the LT to ZT plane of a sampled signal of period T is shown in diagram. The entire jω-axis maps into the unit circle in infinite strips of 2π/T. The left side of the s-plane maps into the inside of the unit circle in the z-plane and the right side maps into the outside of the unit circle. If the unit circle is inside the region of convergence, then the DTFT is obtained from z-transform by substituting z = exp(jωT).

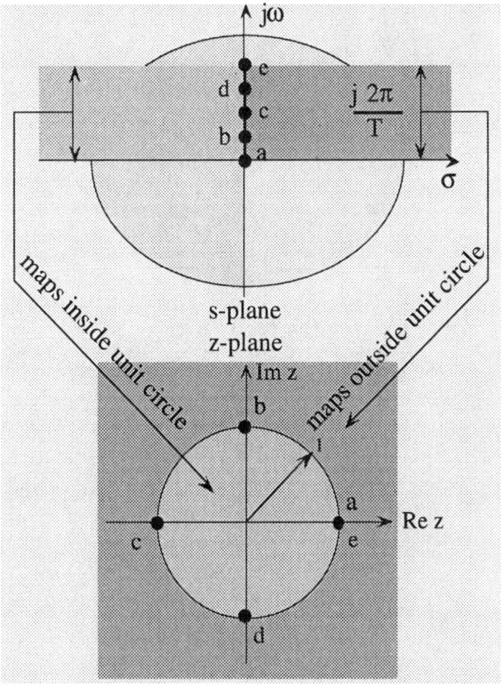

25. PROPERTIES OF BILATERAL z-TRANSFORMS

DEFINITION **$x(n) \overset{z}{\leftrightarrow} X(z)$**

z-Transform:

$$X(z) = \sum_{n=-\infty}^{\infty} y(n)\, z^{-n} : z = re^{j\Omega}$$

$$\text{ROC} \Rightarrow R_x \equiv \{ r_{x1} < r < r_{x2} \}$$

$$Y(z) = \sum_{n=-\infty}^{\infty} y(n)\, z^{-n} : z = re^{j\Omega}$$

$$\text{ROC} \Rightarrow R_y \equiv \{ r_{y1} < r < r_{y2} \}$$

Inverse z-transform:

$$x(n) = \frac{1}{2\pi j} \oint_C X(z)\, z^{n-1} dz$$

$$y(n) = \frac{1}{2\pi j} \oint_C Y(z)\, z^{n-1} dz$$

Frequency convolution:

$$W(z) = \frac{1}{2\pi j} \oint_C X(p) Y\left(\frac{z}{p}\right) p^{-1} dp$$

Time Sequence	Property	z-Transform	ROC
$ax(n) + by(n)$	Linearity	$aX(z) + bY(z)$	$R_x \cap R_y$
$x(n-m)$	Time shifting	$z^{-m}X(z)$	R_x
$x(n-m)$	Time shifting for UZT	$\left[\sum_{k=1}^{m} x(-k) z^{-m+k} + z^{-m}X(z)\right.$	R_x
$z_0^n\, x(n)$	Frequency shifting	$X\left(\frac{z}{z_0}\right)$	$z_0 R_x$

Time Sequence	Property	z-Transform	ROC
$e^{j\Omega_0 n}x(n)$	Exponential multiplication	$X(ze^{-j\Omega_0})$	R_x
$a^n x(n)$	Scaling (a real)	$X\left(\frac{z}{a}\right)$	$\lvert a\rvert R_x$
$x(-n)$	Folding	$X\left(\frac{1}{z}\right)$	$\frac{1}{R_x}$
$x^*(n)$	Conjugate	$X^*\left(\frac{1}{z}\right)$	$\frac{1}{R_x}$
$x(n)*y(n)$	Time convolution	$X(z)Y(z)$	$R_x \cap R_y$
$\left[\sum_{k=-\infty}^{n} x(k) = \sum_{k=0}^{\infty} x(n-k)\right]$	Summation	$\frac{z}{z-1}X(z)$	$\{\lvert z\rvert>1\} \cap R_z$
$x(n)y(n)$	Frequency convolution	$\frac{1}{2\pi j}\oint_C \left[X(p) \times Y\left(\frac{z}{p}\right) p^{-1}\right] dp$	$r_{x1}r_{y1} < r < r_{x2}r_{y2}$
$nx(n)$	Frequency differentiation	$-z\frac{dX(z)}{dz}$	R_x
$x(0)$ (UZT only)	Initial value theorem	$\lim_{z\to 0} X(z)$	
$x(\infty)$ (UZT only)	Final value theorem	$\lim_{z\to 1} (z-1)X(z)$	Poles of $(z-1)X(z)$ lie inside unit circle

26. UNILATERAL z-TRANSFORM PAIRS

DEFINITION: $\mathbf{x(n) \overset{z}{\leftrightarrow} X(z)}$

Unilateral z-transform: $X(z) = \sum_{n=0}^{\infty} x(n) z^{-n}$

ROC: $\{ r > r_{x1} \}$

ROC is always outside the outermost pole

Inverse z-transform: $x(n) = \frac{1}{2\pi j} \oint_C X(z)\, z^{n-1} dz$

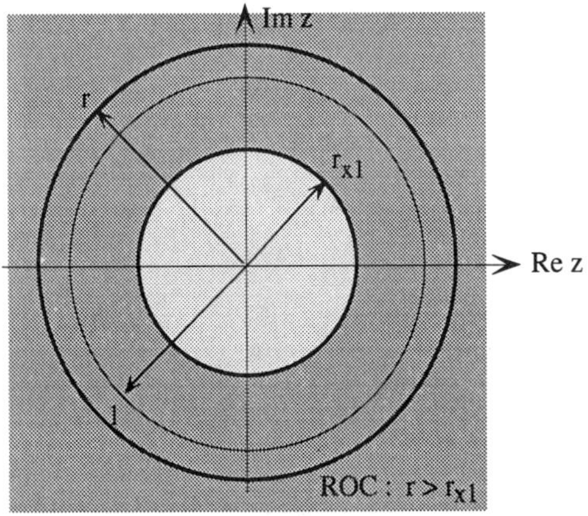

If unit circle is in the region of convergence, $r > r_{x1}$, as shown by the darkly shaded region in the figure above, then DTFT for x(t) exists and is given by substituting $z = e^{j\Omega}$ in X(z).

#	Time Sequence	z-Transform	ROC $|z| >$
1.	$\delta(n)$	1	entire z-plane
2.	$\delta(n-m)$	z^{-m}	0
3.	1	$\frac{z}{z-1}$	1
4.	n	$\frac{z}{(z-1)^2}$	1
5.	n^2	$\frac{z(z+1)}{(z-1)^3}$	1
6.	n^3	$\frac{z(z^2+4z+1)}{(z-1)^4}$	1
7.	a^n	$\frac{z}{z-a}$	$|a|$
8.	na^n	$\frac{az}{(z-a)^2}$	$|a|$
9.	n^2a^n	$\frac{az(z+a)}{(z-a)^3}$	$|a|$
10.	$\frac{a^n}{n!}$	$e^{a/z}$	0
11.	$(n+1)a^n$	$\frac{z^2}{(z-a)^2}$	$|a|$

#	Time Sequence	z-Transform	ROC $|z| >$
12.	$(n+1)(n+2)a^n$	$\frac{z^3}{(z-a)^3}$	$|a|$
13.	$[(n+1)(n+2) \times \ldots (n+m)]a^n$	$\frac{z^{m+1}}{(z-a)^{m+1}}$	$|a|$
14.	$\sin n\omega_0 T$	$\frac{z \sin\omega_0 T}{z^2-2z\cos\omega_0 T+1}$	1
15.	$\cos n\omega_0 T$	$\frac{z(z-\cos\omega_0 T)}{z^2-2z\cos\omega_0 T+1}$	1
16.	$a^n \sin n\omega_0 T$	$\frac{za \sin\omega_0 T}{z^2-2za\cos\omega_0 T+a^2}$	$|a|$
17.	$a^{nT} \sin n\omega_0 T$	$\frac{za^T \sin \omega_0 T}{z^2-2za^T\cos\omega_0 T+a^{2T}}$	$|a|^T$
18.	$a^n \cos n\omega_0 T$	$\frac{z(z-a\cos\omega_0 T)}{z^2-2za\cos\omega_0 T+a^2}$	$|a|$
19.	$a^{nT} \cos n\omega_0 T$	$\frac{z(z-a^T\cos\omega_0 T)}{z^2-2za^T\cos\omega_0 T+a^{2T}}$	$|a|^T$
20.	$e^{-\alpha nT} \sin n\omega_0 T$	$\frac{z\varepsilon^{-\alpha T} z \sin\omega_0 T}{z^2-2ze^{-\alpha T}\cos\omega_0 T+e^{-2\alpha T}}$	$|e^{-\alpha T}|$
21.	$e^{-\alpha nT} \cos n\omega_0 T$	$\frac{z(z-e^{-\alpha T}\cos\omega_0 T)}{z^2-2ze^{-\alpha T}\cos\omega_0 T+e^{-2\alpha T}}$	$|e^{-\alpha T}|$
22.	$e^{-\alpha nT}$	$\frac{z}{z-e^{-\alpha T}}$	$|e^{-\alpha T}|$

#	Time Sequence	z-Transform	ROC $\|z\|$ >
23.	$ne^{-\alpha nT}$	$\dfrac{ze^{-\alpha T}}{\left(z-e^{-\alpha T}\right)^2}$	$\|e^{-\alpha T}\|$
24.	$\dfrac{n(n-1)}{2!}$	$\dfrac{z}{(z-1)^3}$	1
25.	$\dfrac{n(n-1)(n-2)}{3!}$	$\dfrac{z}{(z-1)^4}$	1
26.	$\dfrac{1}{m!}\left[n(n-1)(n-2) \times \cdots (n-m+1)\right]$	$\dfrac{z}{(z-1)^{m+1}}$	1

27. COMPLEX CONVOLUTION (z-Transforms)

We shall demonstrate the important points in complex convolution or frequency convolution of discrete signals. Let x(n) and y(n) be z-transformable discrete signals with z-transforms (ZT) defined by,

$$x(n) \overset{\mathcal{Z}}{\longleftrightarrow} X(z) : \text{ROC} : r_{x1} < r_x^{[1]} < r_{x2}$$
$$y(n) \overset{\mathcal{Z}}{\longleftrightarrow} Y(z) \ : \text{ROC} : r_{y1} < r_y^{[1]} < r_{y2} \tag{1}$$

The ZT of the product w(n) = x(n)y(n) is desired using complex convolution techniques. If the ZT of w(n) is W(z), and the region of convergence for W(z) is $r_{w1} < r_w^{[1]} < r_{w2}$, then we have to determine r_{w1} and r_{w2} in terms of $\{ r_{x1}, r_{y1}\}$ and $\{ r_{x2}, r_{y2}\}$. The existence of X(z) and Y(z) dictates the following relationships:

$$\sum_{n=-\infty}^{\infty} \left| x(n)\, r_x^{-n} \right| < \infty \quad \text{for } r_x > r_{x1}, \ : \ r_x < r_{x2}$$
$$\sum_{n=-\infty}^{\infty} \left| y(n)\, r_y^{-n} \right| < \infty \quad \text{for } r_y > r_{y1} \ : \ r_y < r_{y2} \tag{2}$$

If the z-transform of w(n) = x(n)y(n) is to exist we should have an equation similar to (2),

$$W(z) = \sum_{n=-\infty}^{\infty} \left| w(n)\, r_w^{-n} \right| < \infty$$
$$= \sum_{n=-\infty}^{\infty} \left| x(n)\, r_x^{-n} . y(n)\, r_y^{-n} \right| < \infty \tag{3}$$

Hence, $r_w > (r_{x1}r_{y1})$ and $r_w < (r_{x2}r_{y2})$. Thus, the ROC for W(z) is

$$\text{ROC for } W(z) : (r_{x1}r_{y1}) < r_w < (r_{x2}r_{y2}) \tag{4}$$

and, $r_{w1} = (r_{x1}r_{y1})$ and $r_{w2} = (r_{x2}r_{y2})$. We shall find an expression for W(z) in terms of X(z) and Y(z) as a discrete complex convolution.

---

[1] Subscripts like r_x, r_y and r_w have been added to the ROC, r to indicate the respective variables X, Y, and W.

$$
\begin{aligned}
\mathcal{Z}\{x(n)y(n)\} \\
&= \sum_{n=-\infty}^{\infty} x(n).y(n)\, z^{-n} \\
&= \sum_{n=-\infty}^{\infty} \left[\frac{1}{2\pi j}\oint_C X(p)p^{\,n-1}dp\right] y(n)z^{-n} \\
&= \sum_{n=-\infty}^{\infty} \left[\frac{1}{2pj}\oint_C X(p)\left(\frac{z}{p}\right)^{-n} p^{-n}dp\right] y(n) \qquad (5) \\
&= \frac{1}{2\pi j}\oint_C X(p)\left[\sum_{n=-\infty}^{\infty} y(n)\left(\frac{z}{p}\right)^{-n}\right] p^{-1}dp \\
&= \frac{1}{2\pi j}\oint_C X(p)\, Y\left(\frac{z}{p}\right) p^{-1}dp
\end{aligned}
$$

The contour integrals in eq.(5) have to be evaluated by integrating around a suitably defined contour. This means that the placement of the contour C in the *region of convergence* in the *p-plane* should be carefully determined. The ROCs of X(p) and Y(z/p) in the p-plane are,

$$r_{x1} < r_p < r_{x2}\ ,\ r_{y1} < \frac{r_w}{r_p} < r_{y2}.$$

The multiplication of these two ROCs results in $(r_{x1}\, r_{y1}) < r_w < (r_{x2}\, r_{y2})$ in the z-plane, a result that has been already derived in eq.(4). In summary, to determine the ROC in the p-plane for X(p)Y(z/p) in eq.(5) the following inequalities have to be satisfied:

1. $r_{x1} < r_p < r_{x2}$
2. $r_{y1} < \frac{r_w}{r_p} < r_{y2}$ or $\frac{r_w}{r_{y2}} < r_p < \frac{r_w}{r_{y1}}$

   with r_w taking values as in (4). The ROC for r_p given in (1) and (2) can be determined as the intersection of $\{r_{x1}, r_{x2}\} \cap \left\{\frac{r_w}{r_{y2}}, \frac{r_w}{r_{y1}}\right\}$. This region can be expressed more explicitly by (3) below.

3. $\max\left\{r_{x1}, \frac{r_w}{r_{y2}}\right\} < r_p < \min\left\{r_{x2}, \frac{r_w}{r_{y1}}\right\}$

   with r_w ranging between the products $(r_{x1}r_{y1})$ and $(r_{x2}r_{y2})$ as in (4) below.

4. $(r_{x1}r_{y1}) < r_w < (r_{x2}r_{y2})$

EXAMPLE

Given $x(n) = 4^n u(-n) + u(n)$: $y(n) = u(-n) + 4^{-n} u(n)$, find the ZT of $w(n) = x(n).y(n)$ using complex convolution. Verify by actual multiplication.

The z–transforms X(z) and Y(z) are given by,

$$X(z) = -\frac{4}{z-4} + \frac{z}{z-1} \qquad 1 < r_x < 4 \text{ and } r_{x1} = 1,\ r_{x2} = 4$$

$$Y(z) = -\frac{1}{z-1} + \frac{z}{z-0.25} \quad 0.25 < r_y < 1 \text{ and } r_{y1}=0.25,\ r_{y2}=1 \tag{6}$$

Using condition (3), the ROC for W(z) can be established as $0.25 < r_w < 4$. The regions of convergence for X(z), Y(z) and W(z) are shown in the following figures.

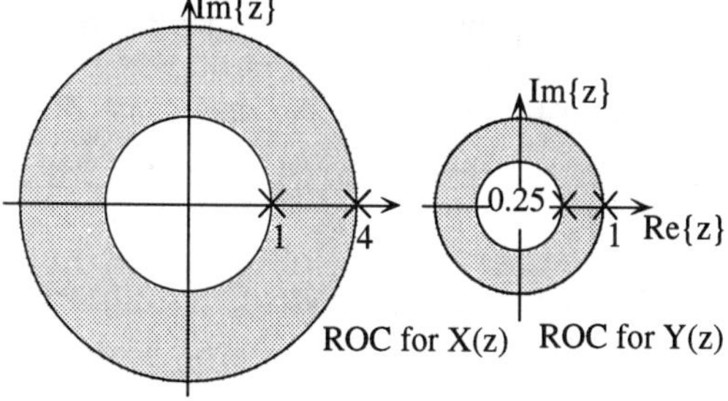

ROC for X(z) ROC for Y(z)

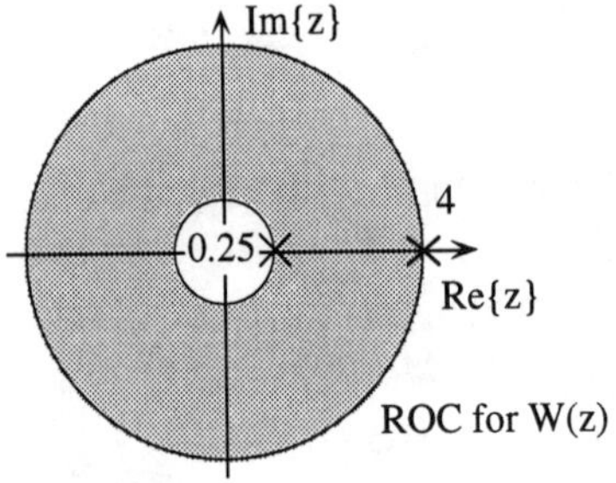

Substituting for X(p) and Y(z/p) in

$$\mathcal{Z}\{x(n)y(n)\} = W(z) = \frac{1}{2\pi j}\oint_C X(p)\, Y\left(\frac{z}{p}\right) p^{-1}\, dp$$

with ROC, $0.25 < r_w < 4$, W(z) can be expressed as:

$$W(z) = \frac{1}{2\pi j}\oint_C \underbrace{\left(-\frac{4}{p-4} + \frac{p}{p-1}\right)}\underbrace{\left(\frac{p}{p-z} - \frac{z}{0.25p-z}\right)} p^{-1} dp \qquad (7)$$

$$\text{ROC: } 1 < r_p < 4 \;:\; \text{ROC: } \frac{r_w}{1} < r_p < \frac{r_w}{0.25}$$

with $r_w = |z|$. We will now plot the poles of the function,

$$\left(-\frac{4}{p-4} + \frac{p}{p-1}\right)\left(\frac{p}{p-z} - \frac{z}{0.25p-z}\right) p^{-1} \qquad (8)$$

in the complex p-plane as shown in the diagram on the next page. The fixed poles are at $p = 0$, $p = 1$ and at $p = 4$. The moving poles are at $p = z$ and at $p = 4z$. Using the inequality,

$$\max\left\{r_{x1}, \frac{r_w}{r_{y2}}\right\} < r_p < \min\left\{r_{x2}, \frac{r_w}{r_{y1}}\right\} \qquad (9)$$

and substituting for r_{x1}, r_{x2}, r_{y1}, r_{y2} from eq.(6) we obtain for this example,

$$\max\{1, r_w\} < r_p < \min\{4, 4r_w\} \qquad (10)$$

where $0.25 < |z| = r_w < 4$. Depending on the ranges of r_w, we obtain two different cases for eq.(10) corresponding to the moving poles at $p = z$ and $p = 4z$.

Case (i) $0.25 < |z| < 1 : 1 < |z| < 4$
ROC for p : $1 < |p| < |4z|$: (shown in figure)
Case (ii) $1 < |z| < 4 : 4 < |4z| < 16$
ROC for p : $|z| < |p| < 4$

The ROC for r_p in the range $(1, |4z|)$ for case (i) is shown in the diagram below. Having located r_p in its ROC we can now evaluate the integral on the contour C. Note that there is no time factor involved. In either of the two cases the contour C contains the three poles at $p = 0$, $p = z$ and $p = 1$. The residues at $p = 0$, $p = z$ and $p = 1$ in contour C are:

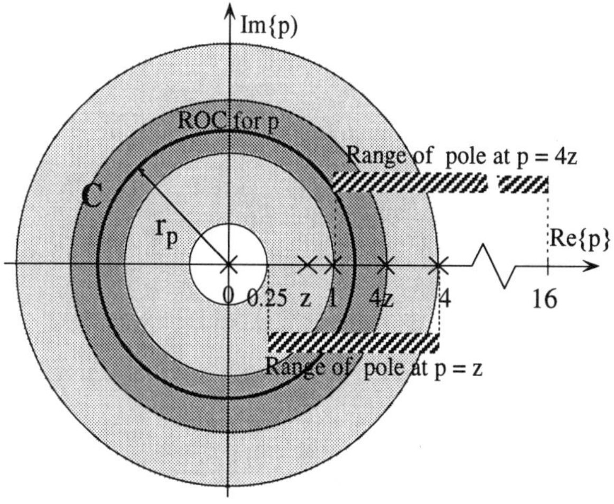

$$K|_{p=0} = 1$$

$$K|_{p=z} = -\frac{4}{z-4} + \frac{z}{z-1}$$

$$K|_{p=1} = -\frac{1}{z-1} + \frac{z}{z-0.25} \qquad (11)$$

We can now obtain W(z) as the sum of residues, that is,

$$W(z) = 1 - \frac{4}{z-4} + \frac{z}{z-1} - \frac{1}{z-1} + \frac{z}{z-0.25} \qquad (12)$$

$$= 2 - \frac{4}{z-4} + \frac{z}{z-0.25}, \; 0.25 < r_w < 4$$

The corresponding time function is given by,

$$w(n) = 2\delta(n) + 4^{-n}u(n) + 4^{n}u(-n) \qquad (13)$$

By direct multiplication of $x(n) = 4^n\, u(-n) + u(n)$ and $y(n) = u(-n) + 4^{-n}\, u(n)$ we obtain,

$w(n) = x(n).y(n) = 2\delta(n) + 4^{-n}u(n) + 4^{n}u(-n)$

that is the same as eq.(13).

28. TRUNCATION WINDOWS

1. Rectangular Window

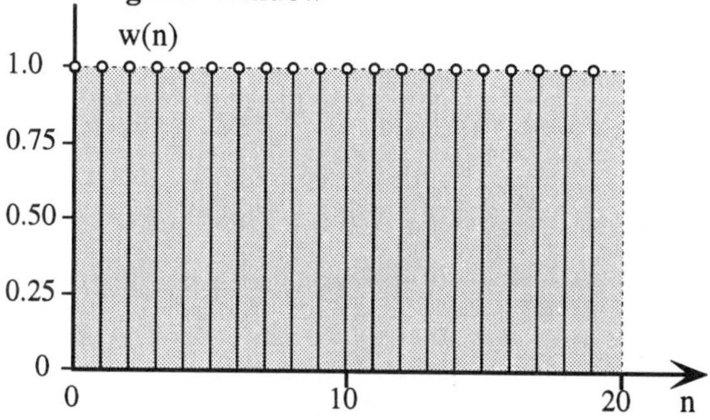

$$w(n) = \begin{cases} 1 & 0 \le n < N-1 \\ 0 & \text{otherwise} \end{cases}$$

2. Bartlett Window (triangular window)

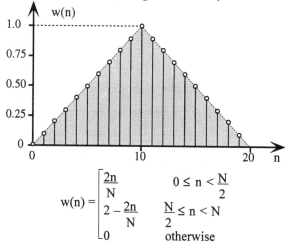

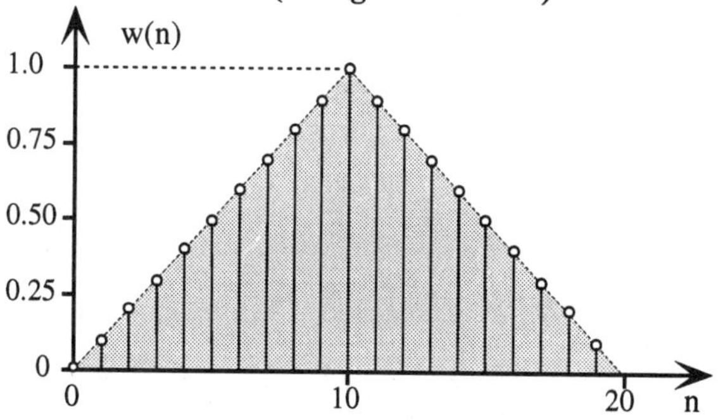

$$w(n) = \begin{cases} \dfrac{2n}{N} & 0 \le n < \dfrac{N}{2} \\ 2 - \dfrac{2n}{N} & \dfrac{N}{2} \le n < N \\ 0 & \text{otherwise} \end{cases}$$

3. Hann Window

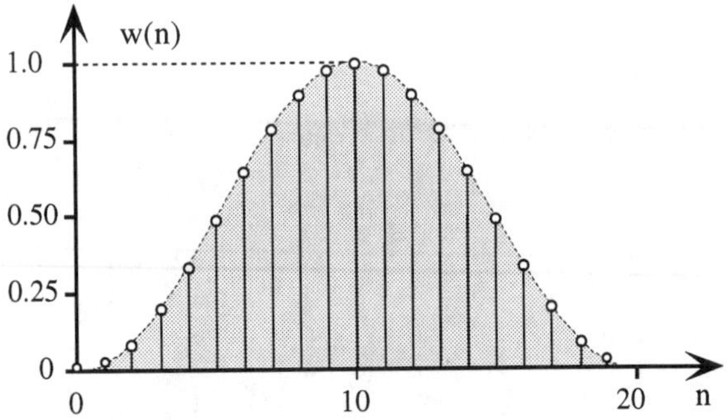

$$w(n) = \begin{cases} 0.5 - 0.5 \cos \dfrac{2\pi n}{N} & 0 \le n < N - 1 \\ 0 & \text{otherwise} \end{cases}$$

4. Hamming Window

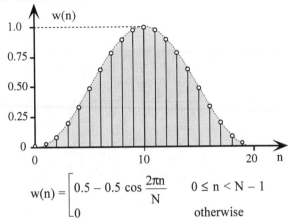

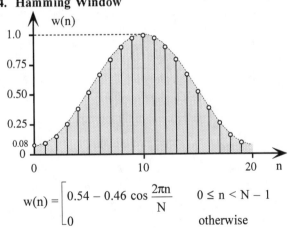

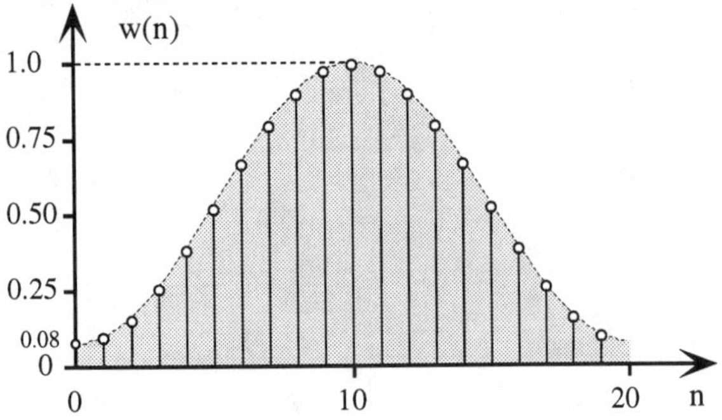

$$w(n) = \begin{cases} 0.54 - 0.46 \cos \dfrac{2\pi n}{N} & 0 \le n < N - 1 \\ 0 & \text{otherwise} \end{cases}$$

5. Blackman Window

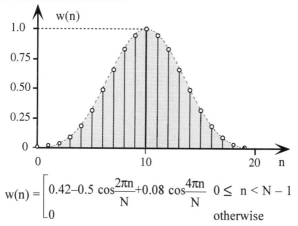

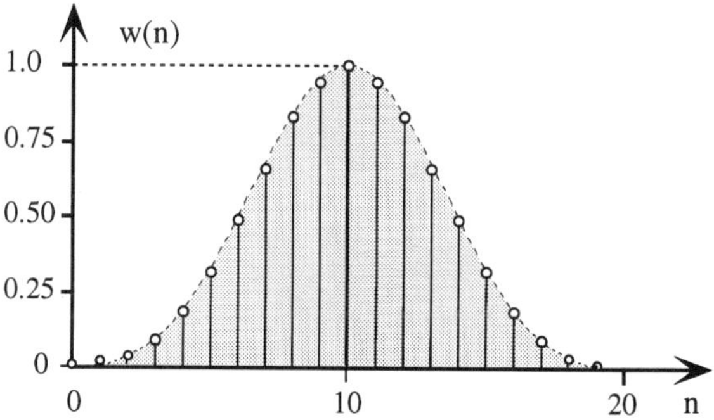

$$w(n) = \begin{cases} 0.42 - 0.5 \cos\dfrac{2\pi n}{N} + 0.08 \cos\dfrac{4\pi n}{N} & 0 \leq n < N - 1 \\ 0 & \text{otherwise} \end{cases}$$

29. LINEAR SPACES

Definitions

A set of n vectors $\{\mathbf{x}_1, \mathbf{x}_2, ..., \mathbf{x}_n\} \in R$, $\mathbf{x}_i = \{x_{i1}, x_{i2}, ..., x_{in}\}^T$ is called an n-dimensional linear space (or affine) if,

1. there exists a vector $z \in R$ called the sum of the pair of vectors $x_i, x_j \in R$ for every i and j, $i \neq j$ denoted by $z = x_i + x_j$.
2. there exists a vector $\mathbf{w} \in R$ called the product of every real number λ and $\mathbf{x}_i \in R$ denoted by $\mathbf{w} = \lambda\mathbf{x}_i$.
3. there exist rules for forming sums and products of elements in R having the following properties:
   a. $\mathbf{x}_i + \mathbf{x}_j = \mathbf{x}_j + \mathbf{x}_i$ for every $\mathbf{x}_i \in R$ (commutativity)
   b. $(\mathbf{x}_i + \mathbf{x}_j) + \mathbf{x}_k = \mathbf{x}_i + (\mathbf{x}_j + \mathbf{x}_k)$ for every $\mathbf{x}_i, \mathbf{x}_j, \mathbf{x}_k \in R$ (associativity)
   c. There exists a zero element $\mathbf{0} \in R$ such that $\mathbf{x}_i + \mathbf{0} = \mathbf{x}_i$ for any $\mathbf{x}_i \in R$ (zero vector)
   d. For every $\mathbf{x}_i \in R$ there exists an inverse element $\mathbf{y}_i \in R$ such that $\mathbf{x}_i + \mathbf{y}_i = 0$ (inverse vector)
   e. $1.\ \mathbf{x}_i = \mathbf{x}_i$ for every $\mathbf{x}_i \in R$
   f. $0.\ \mathbf{x}_i = \mathbf{0}$ for every $\mathbf{x}_i \in R$
   g. $\alpha(\beta\mathbf{x}_i) = (\alpha\beta)\mathbf{x}_i$ for every $\mathbf{x}_i \in R$ and any real α and β (associativity)
   h. $(\alpha + \beta)\mathbf{x}_i = \alpha\mathbf{x}_i + \beta\mathbf{x}_i$ for every $\mathbf{x}_i \in R$ and any real α and β (distributivity)
   i. $\alpha(\mathbf{x}_i + \mathbf{x}_j) = \alpha\mathbf{x}_i + \alpha\mathbf{x}_j$ for every $\mathbf{x}_i \in R$ (distributivity)

Examples

Space R^3: This is the three-dimensional Euclidean space. Each nonzero vector is characterized by a length and direction. The addition of vectors is defined by the parallelogram rule. Multiplication by a constant λ changes the magnitude of the vector and the direction is unchanged if λ is positive and reversed if λ is negative. Two-dimensional and one-dimensional spaces are denoted by R^2 and R^1.

Space T^n: Each element in this space is a vector$\{\mathbf{x}_i, i = 1, \ldots, n\}$ characterized by an ordered set of n real numbers $\{x_{i1}, x_{i2}, \ldots, x_{in}\}$ called components. In this case T^n is called a linear vector space.

Space C(a, b): Each element $f_i(x)$ in this space is characterized by a continuous function defined in the interval [a, b].

A set of vectors in a two-dimensional plane whose initial points are located at the origin of coordinates and the final points lie within the first quadrant does not constitute a linear space because multiplication by –1 takes the vectors away from the first quadrant.

Linear Dependence

Let $\{\mathbf{x}_i, i = 1, \ldots, n\}$ be vectors of the linear space R^n. The vector $\mathbf{y} = \lambda_1\mathbf{x}_1 + \lambda_2\mathbf{x}_2 + \ldots + \lambda_n\mathbf{x}_n$ is called the linear combination of the vectors $\{\mathbf{x}_i, i = 1, \ldots, n\}$. The vectors $\{\mathbf{x}_i, i = 1, \ldots, n\}$ are said to be linearly independent if $\mathbf{y} = 0$ means that $\lambda_1 = \lambda_2 = \ldots = \lambda_n = 0$. If we can find a relationship among the $\{\lambda_i\}$ such that $\mathbf{y} = 0$, then the vectors $\{\mathbf{x}_i, i = 1, \ldots, n\}$ are said to be linearly dependent.

In the R^2 space any two vectors that are not collinear are linearly independent. Any three vectors have to be necessarily linearly dependent. Similarly, in the R^3 space any three vectors that are not coplanar are linearly independent. Any four vectors have to be necessarily linearly dependent.

The linear dependence of a set of vectors $\{\mathbf{x}_i, i = 1, \ldots, n\}$ in the space R_n expressed by the equation,

$$\lambda_1 \mathbf{x}_1 + \lambda_2 \mathbf{x}_2 + \ldots + \lambda_n \mathbf{x}_n = 0 \quad (1)$$

means that the linear relationships expressed by the n equations

$$\begin{bmatrix} \mathbf{x}_1 \, \mathbf{x}_2 \cdots \mathbf{x}_n \end{bmatrix} \begin{bmatrix} \lambda_1 \\ \lambda_2 \\ \vdots \\ \lambda_n \end{bmatrix} = \begin{bmatrix} \lambda_1 x_{11} + \lambda_2 x_{21} + \ldots + \lambda_n x_{n1} \\ \lambda_1 x_{12} + \lambda_2 x_{22} + \ldots + \lambda_n x_{n2} \\ \vdots \\ \lambda_1 x_{1n} + \lambda_2 x_{2n} + \ldots + \lambda_n x_{nn} \end{bmatrix} = 0 \quad (2)$$

hold when not all the constants $\{\lambda_i\}$ are equal to zero. On the other hand if we define a set of n vectors in space R^n given by,

$$\begin{aligned} \mathbf{e}_1 &= (1, 0, 0, \ldots, 0)^T \\ \mathbf{e}_2 &= (0, 1, 0, \ldots, 0)^T \\ &\vdots \\ \mathbf{e}_n &= (0, 0, 0, \ldots, 1)^T \end{aligned} \quad (3)$$

then the set of equations similar to eq.(2)

$$\begin{bmatrix} \mathbf{e}_1 \, \mathbf{e}_2 \cdots \mathbf{e}_n \end{bmatrix} \begin{bmatrix} \lambda_1 \\ \lambda_2 \\ \vdots \\ \lambda_n \end{bmatrix} = \begin{bmatrix} \lambda_1.1 + \lambda_2.0 + \ldots + \lambda_n.0 \\ \lambda_1.0 + \lambda_2.1 + \ldots + \lambda_n.0 \\ \vdots \\ \lambda_1.0 + \lambda_2.0 + \ldots + \lambda_n.1 \end{bmatrix} = 0 \quad (4)$$

has the trivial solution $\lambda_1 = \lambda_2 = \ldots = \lambda_n = 0$. Thus the vectors $\{\mathbf{e}_1, \mathbf{e}_2, \ldots, \mathbf{e}_n\}$ are linearly independent.

Examples:

The set of functions $f_1(x) = \cos^2(x)$, $f_2(x) = \sin^2(x)$ and $f_3(x) = 1$ are not linearly independent since $\cos^2(x) + \sin^2(x) = 1$ and $f_1(x) + f_2(x) - f_3(x) = 0$ holds. On the other hand, a set of functions $\{1, x, x^2, \ldots x^n\}$ are linearly independent because the only solution to the equation,

$$\lambda_1 1 + \lambda_2 x + \lambda_2 x^2 + \ldots + \lambda_n x^n = 0$$

is $\lambda_1 = \lambda_2 = \ldots = \lambda_n = 0$.

Basis vectors

A system of linearly independent vectors $\{\mathbf{e}_1, \mathbf{e}_2, \ldots, \mathbf{e}_n\}$ are called basis vectors for the linear space R^n if for every vector $\mathbf{x}_i \in R^n$ there exists an expansion of the form,

$$\mathbf{x}_i = \lambda_{i1} \mathbf{e}_1 + \lambda_{i2} \mathbf{e}_2 + \ldots + \lambda_{in} \mathbf{e}_n \qquad (5)$$

In this case the coefficients $\{\lambda_{ik}, k = 1, \ldots, n\}$ are called the components of the vector $\mathbf{x}_i$ with respect to the basis vectors $\{\mathbf{e}_i, i = 1, \ldots, n\}$

Examples

Consider in R^2 space the two vectors $\mathbf{e}_1 = (1, 4)$ and $\mathbf{e}_2 = (2, 1)$. Clearly these can be basis vectors in R^2 space because they are linearly independent. We want to express another vector $\mathbf{x} = (-4,5)$ in terms of the basis vectors $\mathbf{e}_1$ and $\mathbf{e}_2$. We express $\mathbf{x}$ as

$$\mathbf{x} = \alpha_1 \mathbf{e}_1 + \alpha_2 \mathbf{e}_2$$

Or, $$(-4, 5) = (\alpha_1, 4\alpha_1) + (2\alpha_2, \alpha_2) \qquad (6)$$

From eq.(6) we have to solve the following two linear equations for α_1, α_2.

$$-4 = \alpha_1 + 2\alpha_2$$
$$5 = 4\alpha_1 + \alpha_2 \qquad (7)$$

Solving for α_1 and α_2 we obtain $\alpha_1 = 2$ and $\alpha_2 = -3$. Thus $\mathbf{x} = 2\mathbf{e_1} - 3\mathbf{e_2}$. This is the expansion for the vector $\mathbf{x}$ in terms of the basis vectors $\mathbf{e_1}$ and $\mathbf{e_2}$.

If we have a different set of basis vectors $\mathbf{d_1}$ and $\mathbf{d_2}$, then we will have different set of coefficients β_1 and β_2 for the vector $\mathbf{x}$. If the two basis vectors are $\mathbf{d_1} = (1, 2)$ and $\mathbf{d_2} = (4, -5)$, then we have similar to eq.(7)

$$-4 = \beta_1 + 4\beta_2$$
$$5 = 2\beta_1 - 5\beta_2 \qquad (8)$$

Solving for β_1 and β_2 we obtain $\beta_1 = 0$ and $\beta_2 = -1$. Thus we have $\mathbf{x} = -\mathbf{d_2}$. It is important to realize that the expansion of any vector is with respect to a given set of basis vectors.

In general, the expansion of any vector $\mathbf{x}$ with respect to an arbitrary set of basis vectors can be quite tedious. However, if the basis vectors form an orthonormal set, then the expansion is comparatively simple.

Dimension of vector space

In a linear space R^n if we can find a set of n linearly independent vectors while every set of (n+1) vectors is linearly dependent, then the number n is known as the dimension of the linear space. Then the space R^n is called an n-dimensional linear space. Hence we can conclude that if there are n basis vectors, then the dimension of the linear space is n. Note that basis vectors need not necessarily be orthonormal. However, if they are orthonormal, then expansion of any vector in terms of the orthonormal basis vectors has desirable properties.

30. BASIC THEORY OF MATRICES

Definitions

A matrix is defined as an array of n vectors of $\{\mathbf{x}_{.j}, j = 1, \ldots, n\}$ with each vector $\mathbf{x}_{.j}$ having m elements $\mathbf{x}_{.j} = \{x_{1j}, x_{2j}, \ldots, x_{mj}\}^T$. These elements can be real or complex. We will assume for the time being that these elements are real and are from a field of real numbers $\mathfrak{F}$. In essence a matrix is a rectangular array of numbers, real or complex, drawn from a field of numbers $\mathfrak{F}$. Thus, the matrix $\mathbf{X}$ is an m × n rectangular array of real or complex numbers and can be expressed as,

$$\mathbf{X} = \left[\mathbf{x}_{.1}\,\mathbf{x}_{.2} \cdots \mathbf{x}_{.j} \cdots \mathbf{x}_{.n}\right]$$

$$= \begin{bmatrix} \mathbf{x}_{.1} & \mathbf{x}_{.2} & \cdots & \mathbf{x}_{.j} & \cdots & \mathbf{x}_{.n} \end{bmatrix} \\ \begin{bmatrix} x_{11} & x_{12} & \cdots & x_{1j} & \cdots & x_{1n} \\ x_{21} & x_{22} & \cdots & x_{2j} & \cdots & x_{2n} \\ \vdots & \vdots & \cdots & \vdots & \cdots & \vdots \\ x_{i1} & x_{i2} & \cdots & x_{ij} & \cdots & x_{in} \\ \vdots & \vdots & \cdots & \vdots & \cdots & \vdots \\ x_{m1} & x_{m2} & \cdots & x_{mj} & \cdots & x_{mn} \end{bmatrix} \qquad (1)$$

Notation: In the double subscript (ij) the first subscript denotes the row and the second subscript denotes the column. Thus an m × n array consists of m rows and n columns. Sometimes the matrix $\mathbf{X}$ is written as $\{x_{ij}\}$.

If the column vectors $\{\mathbf{x}_{.j} = [x_{1j,}\ x_{2j}, \ldots, x_{mj}]^T, j = 1, 2, \ldots n\}$ are linearly independent, then the matrix $\mathbf{X}$ is column independent. On the other hand, if the row vectors denoted by $\{\mathbf{x}_{i.} = [x_{i1}, x_{i2}, \ldots, x_{ij}, \ldots, x_{in}], i = 1, 2, \ldots,$

m}are linearly independent then the matrix **X** is row independent. An arbitrary m × n rectangular matrix can not both be column and row independent unless it is a square matrix of dimension n × n. However, not all square matrices need be both row and column independent.

Matrix Addition

Addition is defined for two matrices **X** and **Y** of similar dimensions. Thus,

$$\mathbf{X} + \mathbf{Y} = \{x_{ij}\} + \{y_{ij}\} = \{x_{ij} + y_{ij}\} \quad (2)$$

The addition rule is commutative, that is **X** + **Y**=**Y** + **X**.

Example

$$\mathbf{X} = \begin{bmatrix} 2\,6\,7 \\ 4\,3\,9 \end{bmatrix} \quad \mathbf{Y} = \begin{bmatrix} 1\,5\,8 \\ 3\,0\,6 \end{bmatrix} \quad \mathbf{Z} = \begin{bmatrix} 2\,7 \\ 6\,3 \end{bmatrix}$$

$$\mathbf{X} + \mathbf{Y} = \begin{bmatrix} 3 & 11 & 15 \\ 7 & 3 & 15 \end{bmatrix}$$

In the above example, **X** + **Y** is defined but **X** + **Z** or **Y** + **Z** is not defined

Matrix Multiplication

Matrix multiplication is a little complex. If we denote the product **XY** = **Z**, then the number of columns of **X** must be equal to the number of rows of **Y**. Otherwise, multiplication is not defined. If **X** is ? × n then **Y** has to be n × ?. In this case the dimension of **Z** is ? × ?. Then **XY** is defined by

$$\mathbf{Z} = \mathbf{XY} = \left\{ \sum_{k=1}^{n} x_{ik}\, y_{kj} \right\} = \{z_{ij}\} \quad (3)$$

Example

$$\mathbf{X} = \begin{bmatrix} 2\,6\,7 \\ 4\,3\,9 \end{bmatrix} \qquad \mathbf{Y} = \begin{bmatrix} 1\,5 \\ 3\,0 \\ 2\,4 \end{bmatrix}$$

$$2 \times 3 \qquad\qquad 3 \times 2$$

$\mathbf{Z} = \mathbf{X\,Y}$

$$= \begin{bmatrix} 2\times1+6\times3+7\times2=34 & 2\times5+6\times0+7\times4=38 \\ 4\times1+3\times3+9\times2=31 & 4\times5+3\times0+9\times4=56 \end{bmatrix}$$

$$2 \times 2$$

In general, matrix multiplication is not commutative, that is, $\mathbf{XY} \neq \mathbf{YX}$. If **X** and **Y** are to commute, then it is necessary that they must be both square and of the same dimension. As an example, consider the following.

$$\mathbf{X} = \begin{bmatrix} 4\,5 \\ 5\,0 \end{bmatrix} \qquad \mathbf{Y} = \begin{bmatrix} -4 & 5 \\ 5 & -8 \end{bmatrix}$$

$$\mathbf{XY} = \begin{bmatrix} 9 & -20 \\ -20 & 25 \end{bmatrix} \qquad \mathbf{YX} = \begin{bmatrix} 9 & -20 \\ -20 & 25 \end{bmatrix}$$

In the above example **X** and **Y** commute. But, in the following example **A** and **B** do not commute.

$$\mathbf{A} = \begin{bmatrix} 4\,5 \\ 5\,0 \end{bmatrix} \qquad \mathbf{B} = \begin{bmatrix} -4 & 5 \\ 5 & -6 \end{bmatrix}$$

$$\mathbf{AB} = \begin{bmatrix} 9 & -10 \\ -20 & 25 \end{bmatrix} \qquad \mathbf{BA} = \begin{bmatrix} 9 & -20 \\ -10 & 25 \end{bmatrix}$$

We usually denote in the product **XY** that **Y** is pre-multiplied by **X**, or **X** is post-multiplied by **Y**. Unlike real numbers, division of a matrix by another matrix is not defined.

Transpose of Matrix

The transpose of a real matrix $\mathbf{X} = \{x_{ij}\}$ is obtained by interchanging the rows and columns and is denoted by $\mathbf{X}^T$

$= \{x_{ij}\}^T = \{x_{ji}\}$. The transpose of a matrix product $(\mathbf{XY})^T$ is given by $\mathbf{Y}^T\mathbf{X}^T$ whereas $(\mathbf{X} + \mathbf{Y})^T = \mathbf{X}^T + \mathbf{Y}^T$.

The conjugate transpose of a complex matrix $\mathbf{Z}$ is denoted by $\mathbf{Z}^H = \mathbf{Z}^{*T}$ and $\mathbf{Z}^H$ is known as the Hermitian transpose of $\mathbf{Z}$.

Examples

$$\mathbf{X} = \begin{bmatrix} 1 & 4 & 5 \\ 2 & 5 & 0 \end{bmatrix} \quad \mathbf{Y} = \begin{bmatrix} 3 & 2 \\ -4 & 5 \\ 5 & -6 \end{bmatrix}$$

$$\mathbf{X}^T = \begin{bmatrix} 1 & 2 \\ 4 & 5 \\ 5 & 0 \end{bmatrix} \quad \mathbf{Y}^T = \begin{bmatrix} 3 & -4 & 5 \\ 2 & 5 & -6 \end{bmatrix}$$

$$(\mathbf{XY})^T = \begin{bmatrix} 12 & -14 \\ -8 & 29 \end{bmatrix} \quad \mathbf{Y}^T\mathbf{X}^T = \begin{bmatrix} 12 & -14 \\ -8 & 29 \end{bmatrix}$$

If $\mathbf{X} = \mathbf{X}^T$, then we say that the matrix $\mathbf{X}$ is symmetric and if $\mathbf{Z} = \mathbf{Z}^H$, then the matrix $\mathbf{Z}$ has Hermitian symmetry or $\mathbf{Z}$ is simply known as a Hermitian matrix.

Determinant of a Matrix

Determinant of a square matrix if it is nonzero is one of the two most important invariants of a matrix. It is usually denoted by det $\mathbf{X}$ or $|\mathbf{X}|$. It is computed as

$$\det \mathbf{X} = \sum_{i=1}^{n} x_{ij}\,\Delta_{ij} \qquad (4)$$

where Δ_{ij} is the cofactor of the matrix $\mathbf{X}$. The cofactor Δ_{ij} is defined by,

$$\Delta_{ij} = (-1)^{i+j} M_{ij} \qquad (5)$$

where M_{ij} is the minor of the matrix $\mathbf{X}$ obtained by taking the determinant after deleting the ith row and the jth column. Eq.(4) is called a cofactor expansion about the ith row. The difference between the cofactor and the minor is the sign as shown in eq.(5).

Examples

The determinant of the following matrix is evaluated as follows.

$$|\mathbf{X}| = \begin{bmatrix} 1 & -1 & 1 & -4 \\ 3 & 4 & -3 & 1 \\ -2 & 3 & 2 & 2 \\ 1 & -1 & 4 & 3 \end{bmatrix} \qquad (6)$$

Expanding in terms of cofactors about the second row the expression for the determinant $|\mathbf{X}|$ is

$$|\mathbf{X}| = -3.\left|\begin{bmatrix} -1 & 1 & -4 \\ 3 & 2 & 2 \\ -1 & 4 & 3 \end{bmatrix}\right| + 4.\left|\begin{bmatrix} 1 & 1 & -4 \\ -2 & 2 & 2 \\ 1 & 4 & 3 \end{bmatrix}\right| + 3.\left|\begin{bmatrix} 1 & -1 & -4 \\ -2 & 3 & 2 \\ 1 & -1 & 3 \end{bmatrix}\right| + 1.\left|\begin{bmatrix} 1 & -1 & 1 \\ -2 & 3 & 2 \\ 1 & -1 & 4 \end{bmatrix}\right| \qquad (7)$$

Each one of the determinants in eq.(7) can again be expanded in terms of minors as shown by the expansion of the first term.

$$\begin{aligned} -3.\left|\begin{bmatrix} -1 & 1 & -4 \\ 3 & 2 & 2 \\ -1 & 4 & 3 \end{bmatrix}\right| &= -3.\left[-1.\left|\begin{bmatrix} 2 & 2 \\ 4 & 3 \end{bmatrix}\right| - 1.\left|\begin{bmatrix} 3 & 2 \\ -1 & 3 \end{bmatrix}\right| - 4.\left|\begin{bmatrix} 3 & 2 \\ -1 & 4 \end{bmatrix}\right|\right] \\ &= -3.-65 = 195 \end{aligned} \qquad (8)$$

Evaluating the determinant we have $|\mathbf{X}| = -3 \times (-65) + 4 \times 46 + 3 \times 7 + 3 = 403$.

If the determinant is zero, we call that matrix a singular matrix or "simply degenerate." If the determinants of the first and higher order minors are also zero, then we call the matrix "multiply degenerate."

Properties of Determinants (X and Y nonsingular)

1. det **X** = det $\mathbf{X}^T$.
   The following properties (2, 3, 4, 5, 6) are very useful in the manipulation of matrices.
2. If any row (or column) of **X** is multiplied by a constant a to yield a new matrix **Y**, then det **X** = a.det **Y**.
3. If any rows (or columns) of **X** are proportional to one another, then the rows or columns are linearly dependent and det **X** = 0.
4. If any two rows (or columns) of **X** are interchanged an odd number of times to yield **Y**, then det **Y** = –det **X**. If they are interchanged an even number of times to yield **Z**, then det **Z** = det **X**.
5. If any row (or column) is modified by adding to it α times the corresponding elements of another row (or column) to yield **Y**, then det **Y** = det **X**.
6. The determinant of a triangular matrix **X** is the product of its diagonal terms:
   det **X** = $x_{11}.x_{22}.\ldots.x_{nn}$.
7. If the matrices **X** and **Y** are both n × n, then det (**XY**) = (det **X**)(det **Y**).
8. det (α**X**) = α^n det **X**.
9. Even if **X** and **Y** are n × n, det(**X** + **Y**) ≠ det **X** + det **Y** in general.

Trace of a Matrix

The trace of a square matrix Tr(**X**) is the second important invariant of a matrix. The trace is the sum of the diagonal terms of a matrix. The trace of the matrix given in eq.(6) is 1 + 4 + 2 + 3 = 10

Rank of a Matrix

The column rank of a rectangular matrix is the number of linearly independent column vectors. Similarly, the row rank of a matrix is the number of linearly independent rows. If the column rank and the row rank are the same, then we call that number the rank of the matrix. Thus a nonsingular square matrix has the column rank equal to the row rank and the rank is the dimension of the matrix. In general, the rank of any m × n matrix is also defined as the greatest possible nonzero determinant formed by taking 2, 3, …, n or m elements at a time. Forming all possible combinations of determinants from any given matrix is a difficult task. Finding a rank is not always easy, but some methods are easier than others. We show this by means of a running example for finding the ranks and the determinants of nonsingular and singular matrices.

Example (Full Rank)

Find the rank and determinant of the following matrix.

$$\mathbf{X} = \begin{bmatrix} 1 & 3 & 5 & 4 & 1 \\ 2 & -1 & -2 & -3 & 4 \\ -1 & 4 & -4 & 2 & -5 \\ 3 & 2 & 1 & 0 & 3 \\ 4 & 0 & -3 & -1 & 2 \end{bmatrix} \tag{9}$$

Method 1 (Manipulation of columns of the matrix)

The idea of this method is to manipulate the matrix by adding corresponding columns multiplied by a constant (property 5) to yield a triangular matrix. The determinant is obtained by taking the product of the diagonal terms (property 6) of the resulting matrix. In the process of triangularization, if some of the rows or columns are zeros, then the rank of the matrix is reduced by the number of rows or columns that are zero. The product of the remaining terms in the diagonal gives the determinant of the reduced order nonsingular matrix.

In the matrix of eq.(9) the first column remains as it is. We subtract 3 times the first column from the second column, 5 times the first column from the third column, 4 times the first column from the fourth column and the first column from the fifth column. This results in the last 4 terms in the first row except the first being 0.

$$\mathbf{X} = \begin{bmatrix} 1 & 3-3 & 5-5 & 4-4 & 1-1 \\ 2 & -1-6 & -2-10 & -3-8 & 4-2 \\ -1 & 4+3 & -4+5 & 2+4 & -5+1 \\ 3 & 2-9 & 1-15 & 0-12 & 3-3 \\ 4 & 0-12 & -3-20 & -1-16 & 2-4 \end{bmatrix}$$

$$= \begin{bmatrix} 1 & 0 & 0 & 0 & 0 \\ 2 & -7 & -12 & -11 & 2 \\ -1 & 7 & 1 & 6 & -4 \\ 3 & -7 & -14 & -12 & 0 \\ 4 & -12 & -23 & -17 & -2 \end{bmatrix} \qquad (10)$$

The next step is to make the last three terms in the second row to 0. To this effect we subtract (12/7) times the second column from the third column, (11/7) times the second column from the fourth column and add (2/7) times the second column to the fifth column. This results in the following equation.

$$\begin{matrix} 1 & 0 & 0 & 0 & 0 \\ 2 & -7 & -12+\frac{12}{7}.7 & -11+\frac{11}{7}.7 & 2-\frac{2}{7}.7 \\ -1 & 7 & 1-\frac{12}{7}.7 & 6-\frac{11}{7}.7 & -4+\frac{2}{7}.7 \\ 3 & -7 & -14+\frac{12}{7}.7 & -12+\frac{11}{7}.7 & 0-\frac{2}{7}.7 \\ 4 & -12 & -23+\frac{12}{7}.7 & -17+\frac{11}{7}.12 & -2+\frac{2}{7}.12 \end{matrix}$$

$$= \begin{bmatrix} 1 & 0 & 0 & 0 & 0 \\ 2 & -7 & 0 & 0 & 0 \\ -1 & 7 & -11 & -5 & -2 \\ 3 & -7 & -2 & -1 & -2 \\ 4 & -12 & -\frac{17}{7} & \frac{13}{7} & -\frac{38}{7} \end{bmatrix} \qquad (11)$$

The succeeding steps are shown in the following sequence of matrices.

$$\mathbf{X} = \begin{bmatrix} 1 & 0 & 0 & 0 & 0 \\ 2 & -7 & 0 & 0 & 0 \\ -1 & 7 & -11 & -5 & -2 \\ 3 & -7 & -2 & -1 & -2 \\ 4 & -12 & -\frac{17}{7} & \frac{13}{7} & -\frac{38}{7} \end{bmatrix} = \begin{bmatrix} 1 & 0 & 0 & 0 & 0 \\ 2 & -7 & 0 & 0 & 0 \\ -1 & 7 & -11 & 0 & 0 \\ 3 & -7 & -2 & -\frac{1}{11} & -\frac{18}{11} \\ 4 & -12 & -\frac{17}{7} & \frac{228}{77} & -\frac{384}{77} \end{bmatrix}$$

Or,

$$\mathbf{X} = \begin{bmatrix} 1 & 0 & 0 & 0 & 0 \\ 2 & -7 & 0 & 0 & 0 \\ -1 & 7 & -11 & 0 & 0 \\ 3 & -7 & -2 & -\frac{1}{11} & 0 \\ 4 & -12 & -\frac{17}{7} & \frac{228}{77} & -\frac{408}{7} \end{bmatrix} \qquad (12)$$

Since five steps were needed for this reduction we conclude that the rank of the matrix is 5 and the determinant is given by,

$$1 \times -7 \times -11 \times -\frac{1}{11} \times -\frac{408}{7} = 408$$

Method 2 (Using pivotal 2 × 2 determinants)

This method involves reducing the order of an $n \times n$ matrix $\mathbf{X}$ by sequentially reducing the order using the first element (x_{11}) as the pivot and taking the 2×2 determinants with the pivot. This is shown in the following example.

$$\mathbf{X} = \begin{bmatrix} x_{11} & x_{12} & x_{13} & \cdots & x_{1k} & \cdots & x_{1n} \\ x_{21} & x_{22} & x_{23} & \cdots & x_{2k} & \cdots & x_{2n} \\ x_{31} & x_{32} & x_{33} & \cdots & x_{3k} & \cdots & x_{3n} \\ \vdots & \vdots & \vdots & \vdots & \vdots & \vdots & \vdots \\ x_{j1} & xa_{j2} & x_{j3} & \cdots & x_{jk} & \cdots & x_{jn} \\ \vdots & \vdots & \vdots & \vdots & \vdots & \vdots & \vdots \\ x_{n1} & x_{n2} & x_{n3} & \cdots & x_{nk} & \cdots & x_{nn} \end{bmatrix}$$

$$
= x_{11} \begin{bmatrix} 1 & x_{12}/x_{11} & x_{13}/x_{11} & \cdots & x_{1k}/x_{11} & \cdots & x_{1n}/x_{11} \\ x_{21} & x_{22} & x_{23} & \cdots & x_{2k} & \cdots & x_{2n} \\ x_{31} & x_{32} & x_{33} & \cdots & x_{3k} & \cdots & x_{3n} \\ \vdots & \vdots & \vdots & \vdots & \vdots & \vdots & \vdots \\ x_{j1} & x_{j2} & x_{j3} & \cdots & x_{jk} & \cdots & x_{jn} \\ \vdots & \vdots & \vdots & \vdots & \vdots & \vdots & \vdots \\ x_{n1} & x_{n2} & x_{n3} & \cdots & x_{nk} & \cdots & x_{nn} \end{bmatrix} \quad (13)
$$

We determine the rank and, if singular, the maximum determinant using the following procedure.

Step 1

Using property (2), remove x_{11} outside the matrix as shown in eq.(13). We form an (n–1)×(n–1) matrix $\mathbf{X}_1$ from 2 × 2 determinants of the form $\{x_{11}.x_{jk} - x_{j1}.x_{1k}$: j, k = 1, …, n }. The resulting (n–1)×(n–1) matrix is shown.

$$
\mathbf{X}_1 = x_{11} \begin{bmatrix} x_{22} - x_{21}\dfrac{x_{12}}{x_{11}} & x_{23} - x_{21}\dfrac{x_{13}}{x_{11}} & \cdots \\ x_{32} - x_{31}\dfrac{x_{12}}{x_{11}} & x_{33} - x_{31}\dfrac{x_{13}}{x_{11}} & \cdots \\ \vdots & \vdots & \vdots \\ x_{j2} - x_{j1}\dfrac{x_{12}}{x_{11}} & x_{j3} - x_{j1}\dfrac{x_{13}}{x_{11}} & \cdots \\ \vdots & \vdots & \vdots \\ x_{n2} - x_{n1}\dfrac{x_{12}}{x_{11}} & x_{n3} - x_{n1}\dfrac{x_{13}}{x_{11}} & \cdots \end{bmatrix}
$$

$$
\left.\begin{matrix}
x_{2k}-x_{21}\frac{x_{1k}}{x_{11}} & \cdots & x_{2n}-x_{21}\frac{x_{1n}}{x_{11}} \\
x_{3k}-x_{31}\frac{x_{1k}}{x_{11}} & \cdots & x_{3n}-x_{31}\frac{x_{1n}}{x_{11}} \\
\vdots & \vdots & \vdots \\
x_{jk}-x_{j1}\frac{x_{1k}}{x_{11}} & \cdots & x_{jn}-x_{j1}\frac{x_{1n}}{x_{11}} \\
\vdots & \vdots & \vdots \\
x_{nk}-x_{n1}\frac{x_{1k}}{x_{11}} & \cdots & x_{nn}-x_{n1}\frac{x_{1n}}{x_{11}}
\end{matrix}\right] \quad (14)
$$

Step 2

Normalize the first element of $\mathbf{X}_1$ by dividing its first row by $x_{22}-x_{21}\frac{x_{12}}{x_{11}}$ once again, as shown below.

$$
\mathbf{X}_1 = x_{11}\left(x_{22}-x_{21}\frac{x_{12}}{x_{11}}\right)\left[\begin{matrix}
1 & \dfrac{x_{23}-x_{21}\dfrac{x_{13}}{x_{11}}}{x_{22}-x_{21}\dfrac{x_{12}}{x_{11}}} \\
x_{32}-x_{31}\frac{x_{12}}{x_{11}} & x_{33}-x_{31}\frac{x_{13}}{x_{11}} \\
\vdots & \vdots \\
x_{j2}-x_{j1}\frac{x_{12}}{x_{11}} & x_{j3}-x_{j1}\frac{x_{13}}{x_{11}} \\
\vdots & \vdots \\
x_{n2}-x_{n1}\frac{x_{12}}{x_{11}} & x_{n3}-x_{n1}\frac{x_{13}}{x_{11}}
\end{matrix}\right.
$$

$$\left.\begin{matrix} \cdots & \dfrac{x_{2k} - x_{21}\dfrac{x_{1k}}{x_{11}}}{x_{22} - x_{21}\dfrac{x_{12}}{x_{11}}} & \cdots & \dfrac{x_{2n} - x_{21}\dfrac{x_{1n}}{x_{11}}}{x_{22} - x_{21}\dfrac{x_{12}}{x_{11}}} \\ \cdots & x_{3k} - x_{31}\dfrac{x_{1k}}{x_{11}} & \cdots & x_{3n} - x_{31}\dfrac{x_{1n}}{x_{11}} \\ \vdots & \vdots & \vdots & \vdots \\ \cdots & x_{jk} - x_{j1}\dfrac{x_{1k}}{x_{11}} & \cdots & x_{jn} - x_{j1}\dfrac{x_{1n}}{x_{11}} \\ \vdots & \vdots & \vdots & \vdots \\ \cdots & x_{nk} - x_{n1}\dfrac{x_{1k}}{x_{11}} & \cdots & x_{nn} - x_{n1}\dfrac{x_{1n}}{x_{11}} \end{matrix}\right] \quad (15)$$

Step 3

Form a new $(n - 2) \times (n - 2)$ matrix $\mathbf{X_2}$ in an exactly analogous fashion by forming 2×2 determinants as in step 1.

Step 4

Continue this process of reducing the order of the matrix until, either we end up with a scalar (1×1 matrix) or a matrix whose elements are all zero.

Step 5

If the termination ended in a scalar, then the number of multiplications gives the rank of the matrix and the resulting product is the determinant. If the termination ended in a matrix of zeros, then the number of multiplications before the matrix of zeros gives the rank and the product gives the maximum determinant.

We shall now use this method to find the rank and determinant of the full rank matrix of eq.(9).

Example (Full Rank)

$$\mathbf{X} = \begin{bmatrix} 1 & 3 & 5 & 4 & 1 \\ 2 & -1 & -2 & -3 & 4 \\ -1 & 4 & -4 & 2 & -5 \\ 3 & 2 & 1 & 0 & 3 \\ 4 & 0 & -3 & -1 & 2 \end{bmatrix} \qquad (16)$$

We shall indicate the sequence of matrices corresponding to the five steps discussed earlier. Since the leading number is 1 we directly proceed to form the matrix $\mathbf{X}_1$ with 4×4 determinants as indicated in **step 1**.

$$\mathbf{X}_1 = 1.\begin{bmatrix} (-1)\,2.3 & (-2)-2.5 & (-3)-2.4 & 4-2.1 \\ 4-(-1).3 & (-4)-(-1).5 & 2-(-1).4 & (-5)-(-1).1 \\ 2-3.3 & 1-3.5 & 0-3.4 & 3-3.1 \\ 0-4.3 & (-3)-4.5 & (-1)-4.4 & 2-4.1 \end{bmatrix}$$

$$= 1.\begin{bmatrix} -7 & -12 & -11 & 2 \\ 7 & 1 & 6 & -4 \\ -7 & -14 & -12 & 0 \\ -12 & -23 & -17 & -2 \end{bmatrix} \qquad (17)$$

Since the leading number in $\mathbf{X}_1$ is –7, we normalize the leading number and the resulting matrix is,

$$\mathbf{X}_1 = 1.\,(-7).\begin{bmatrix} 1 & \frac{12}{7} & \frac{11}{7} & \frac{2}{7} \\ 7 & 1 & 6 & -4 \\ -7 & -14 & -12 & 0 \\ -12 & -23 & -17 & -2 \end{bmatrix} \qquad (18)$$

We form a new 3×3 matrix $\mathbf{X_2}$ of **step** 2 as follows:

$$\mathbf{X}_2 = (-7).\begin{bmatrix} 1-7.\frac{12}{7} & 6-7.\frac{11}{7} & -4-7.\left(-\frac{2}{7}\right) \\ -14-\left(-7.\frac{12}{7}\right) & -12-\left(-7.\frac{11}{7}\right) & -\left(-7.\left(-\frac{2}{7}\right)\right) \\ -23-\left(-12.\frac{12}{7}\right) & -17-\left(-12.\frac{11}{7}\right) & -2-\left(-12.\left(-\frac{2}{7}\right)\right) \end{bmatrix}$$

$$= \begin{bmatrix} -11 & -5 & -2 \\ -2 & -1 & -2 \\ -\frac{17}{7} & \frac{13}{7} & -\frac{38}{7} \end{bmatrix} \tag{19}$$

Since the leading number in $\mathbf{X_2}$ is -11, we normalize this matrix as follows:

$$\mathbf{X}_2 = 1.\ (-7).(-11)\begin{bmatrix} 1 & \frac{5}{11} & \frac{2}{11} \\ -2 & -1 & -2 \\ -\frac{17}{7} & \frac{13}{7} & -\frac{38}{7} \end{bmatrix} \tag{20}$$

We now form a 2×2 matrix $\mathbf{X_3}$ given below with the corresponding normalization:

$$\mathbf{X}_3 = (-7).(-11)\begin{bmatrix} -1-(-2).\frac{5}{11} & -2-(-2).\frac{2}{11} \\ \frac{13}{7}-\left(-\frac{17}{7}\right)\frac{5}{11} & -\frac{38}{7}-\left(-\frac{17}{7}\right)\frac{2}{11} \end{bmatrix}$$

$$= 1.\ (-7).(-11)\begin{bmatrix} \frac{-1}{11} & \frac{-18}{11} \\ \frac{228}{77} & \frac{-384}{77} \end{bmatrix}$$

$$= 1\,.\,(-7).(-11).\left(-\frac{1}{11}\right)\begin{bmatrix} 1 & 18 \\ \frac{228}{77} & \frac{-384}{77} \end{bmatrix} \qquad (21)$$

The final step is to take the determinant of the 2 × 2 matrix in eq.(21) and the result is,

$$\mathbf{X}_4 = 1\,.\,(-7).(-11).\left(-\frac{1}{11}\right)\left(-\frac{408}{7}\right).$$

Since there are 5 multiplications we conclude that the rank is 5 and the determinant of $\mathbf{X} = \mathbf{X}_4 = 408$, confirming the answer in the previous example.

Example (Rank Deficient)

We shall do a second example where a 5 × 5 matrix is not of full rank, that is, less than 5. The following matrix is given. We are again required to determine the rank and the maximum determinant for this matrix.

$$\mathbf{X} = \begin{bmatrix} 6 & 11 & 5 & 4 & 1 \\ 5 & 3 & -2 & -3 & 4 \\ -8 & -12 & -4 & 2 & -5 \\ 6 & 7 & 1 & 0 & 3 \\ 3 & 0 & -3 & -1 & 2 \end{bmatrix} = 6\begin{bmatrix} 1 & \frac{11}{6} & \frac{5}{6} & \frac{2}{3} & \frac{1}{6} \\ 5 & 3 & -2 & -3 & 4 \\ -8 & -12 & -4 & 2 & -5 \\ 6 & 7 & 1 & 0 & 3 \\ 3 & 0 & -3 & -1 & 2 \end{bmatrix} \qquad (22)$$

$$\mathbf{X}_1 = 6\begin{bmatrix} -\frac{37}{6} & -\frac{37}{6} & -\frac{19}{6} & \frac{19}{6} \\ \frac{8}{3} & \frac{8}{3} & \frac{22}{3} & -\frac{11}{3} \\ -4 & -4 & -4 & 2 \\ -\frac{11}{2} & -\frac{11}{2} & -3 & \frac{3}{2} \end{bmatrix} = 6\left(-\frac{37}{6}\right).\begin{bmatrix} 1 & 1 & \frac{38}{37} & -\frac{19}{37} \\ \frac{8}{3} & \frac{8}{3} & \frac{22}{3} & -\frac{11}{3} \\ -4 & -4 & -4 & 2 \\ -\frac{11}{2} & -\frac{11}{2} & -3 & \frac{3}{2} \end{bmatrix} \qquad (23)$$

$$\mathbf{X}_2=6.\left(\frac{-37}{6}\right).\begin{bmatrix} 0 & \frac{170}{37} & \frac{-85}{37} \\ 0 & \frac{4}{37} & \frac{-2}{37} \\ 0 & \frac{98}{37} & \frac{-49}{37} \end{bmatrix} =6.\left(\frac{-37}{6}\right).\begin{bmatrix} \frac{170}{37} & \frac{-85}{37} & 0 \\ \frac{4}{37} & \frac{-2}{37} & 0 \\ \frac{98}{37} & \frac{-49}{37} & 0 \end{bmatrix}$$

$$= 6.\left(\frac{-37}{6}\right).\left(\frac{170}{37}\right).\begin{bmatrix} 1 & \frac{-1}{2} & 0 \\ \frac{4}{37} & \frac{-2}{37} & 0 \\ \frac{98}{37} & \frac{-49}{37} & 0 \end{bmatrix} \tag{24}$$

Finally, $\mathbf{X}_3 = 6.\left(\frac{-37}{6}\right).\left(\frac{170}{37}\right).\begin{bmatrix} 0 & 0 \\ 0 & 0 \end{bmatrix}$ (25)

In this case we see that this process of reduction has resulted in a premature termination resulting in a 2×2 matrix of zeros. There are 3 multiplications and hence the rank of this matrix is 3 and the maximum possible determinant is –170.

Inverse of a Matrix

Since division by a matrix is not defined we define the inverse of a nonsingular matrix. The inverse of a nonsingular matrix **X** is the matrix $\mathbf{X}^{-1}$ such that $\mathbf{X}\mathbf{X}^{-1} = \mathbf{X}^{-1}\mathbf{X} = \mathbf{I}$, an identity matrix. Thus inverse matrices commute with their direct counterparts. The cofactor expansion of the determinant of the matrix **X** from eq.(4) is

$$\det \mathbf{X} = \sum_{i=1}^{n} x_{ij}\,\Delta_{ij} \tag{4}$$

Eq.(4) can be rewritten as

$$\det \mathbf{X} = \sum_{i=1}^{n} x_{ij}\,\Delta_{ij} = \sum_{i=1}^{n} x_{ij}\,\Delta_{ik} = \begin{cases} \det \mathbf{X}, & k = j \\ 0, & k \neq j \end{cases} \tag{26}$$

In eq.(26), when k = j, it is det **X** by eq.(4), but when k ≠ j, then we have cofactor expansion about the ith column but with cofactors obtained from the kth column and not the ith column. Thus the determinant will contain two identical columns and hence from property 3, det **X** = 0. We divide eq.(26) by det **X** and obtain

$$\sum_{i=1}^{n} \left(\frac{\Delta_{ik}}{\det \mathbf{X}}\right) x_{ij} = \delta_{kj}\ , \begin{cases} 1 \leq k \leq n \\ 1 \leq j \leq n \end{cases} \qquad (27)$$

This is a scalar statement of the matrix equation $\mathbf{X}^{-1}\mathbf{X} = \mathbf{I}$ and the expression for the inverse is given by,

$$\mathbf{X}^{-1} = \{\alpha_{ki}\} = \left\{\frac{\Delta_{ik}}{\det \mathbf{X}}\right\} = \frac{1}{\det \mathbf{X}} \begin{bmatrix} \Delta_{11} \Delta_{21} \cdots \Delta_{n1} \\ \Delta_{12} \Delta_{22} \cdots \Delta_{n2} \\ \vdots \quad \vdots \quad \vdots \quad \vdots \\ \Delta_{1n} \Delta_{2n} \cdots \Delta_{nn} \end{bmatrix} \qquad (28)$$

Note that the cofactors are in the reversed order from the ordering of the elements of the matrix **X**.

The matrix $\{\Delta_{ji}\}$ of the cofactors is sometimes called the *adjoint* matrix. Finding the inverse using eq.(28) is not only very inefficient but also gives erroneous results if the spread of eigenvalues is large. It is only of theoretical interest since it gives an explicit expression for the inverse. In Section 32 we will discuss a more efficient method for matrix inversion.

Solution of Linear Equations

We can use matrix analysis to solve for a set of linear equations of the form **Ax** = **y** as shown below.

$$\begin{bmatrix} a_{11} & a_{12} & \cdots & a_{1j} & \cdots & a_{1n} \\ a_{21} & a_{22} & \cdots & a_{2j} & \cdots & a_{2n} \\ \vdots & \vdots & \vdots & \vdots & \vdots & \vdots \\ a_{i1} & a_{i2} & \cdots & a_{ij} & \cdots & a_{in} \\ \vdots & \vdots & \vdots & \vdots & \vdots & \vdots \\ a_{n1} & a_{n2} & \cdots & a_{nj} & \cdots & a_{nn} \end{bmatrix} \begin{bmatrix} x_1 \\ x_2 \\ \vdots \\ x_j \\ \vdots \\ x_n \end{bmatrix} = \begin{bmatrix} y_1 \\ y_2 \\ \vdots \\ y_i \\ \vdots \\ y_n \end{bmatrix}$$

We shall investigate the case where **A** is a nonsingular $n \times n$ matrix and **x** and **y** are $n \times 1$ vectors. We want to solve for **x** in terms of **y**. Since **A** is invertible we can write the solution as

$$\mathbf{x} = \mathbf{A}^{-1}\,\mathbf{y} \tag{29}$$

Eq.(29) can be expanded and written as,

$$\begin{bmatrix} x_1 \\ x_2 \\ \vdots \\ x_i \\ \vdots \\ x_n \end{bmatrix} = \frac{1}{\det \mathbf{A}} \begin{bmatrix} \Delta_{11} & \Delta_{21} & \cdots & \Delta_{j1} & \cdots & \Delta_{n1} \\ \Delta_{12} & \Delta_{22} & \cdots & \Delta_{j2} & \cdots & \Delta_{n2} \\ \vdots & \vdots & \vdots & \vdots & \vdots & \vdots \\ \Delta_{1i} & \Delta_{2i} & \cdots & \Delta_{ji} & \cdots & \Delta_{ni} \\ \vdots & \vdots & \vdots & \vdots & \vdots & \vdots \\ \Delta_{1n} & \Delta_{2n} & \cdots & \Delta_{jn} & \cdots & \Delta_{nn} \end{bmatrix} \begin{bmatrix} y_1 \\ y_2 \\ \vdots \\ y_j \\ \vdots \\ y_n \end{bmatrix}$$

$$= \frac{1}{\det \mathbf{A}} \sum_{j=1}^{n} \begin{bmatrix} \Delta_{j1}\, y_j \\ \Delta_{j2}\, y_j \\ \vdots \\ \Delta_{ji}\, y_j \\ \vdots \\ \Delta_{jn}\, y_j \end{bmatrix} \tag{30}$$

Note that the order of the cofactors is inverted from that of the original matrix **A**, that is, if a_{ij} is an element in the ith row and jth column Δ_{ij} is in the jth row and ith column.

Example

We shall now solve the following set of linear equations.

$$\begin{aligned}
\frac{8}{3}x_1 - 2x_2 - x_3 + \frac{1}{3}x_4 &= 2 \\
-\frac{7}{3}x_1 + 3x_2 + \frac{7}{3}x_3 - \frac{4}{3}x_4 &= 7 \\
-2x_1 + 4x_2 + \frac{1}{3}x_3 - \frac{2}{3}x_4 &= 6 \\
x_1 - 2x_2 - x_3 + 2x_4 &= -8
\end{aligned} \tag{31}$$

The above set can expressed in a matrix form as follows.

$$\begin{bmatrix} \frac{8}{3} & -2 & -1 & \frac{1}{3} \\ -\frac{7}{3} & 3 & \frac{7}{3} & -\frac{4}{3} \\ -2 & 4 & \frac{1}{3} & -\frac{2}{3} \\ 1 & -2 & -1 & 2 \end{bmatrix} \begin{bmatrix} x_1 \\ x_2 \\ x_3 \\ x_4 \end{bmatrix} = \begin{bmatrix} 2 \\ 7 \\ 6 \\ -8 \end{bmatrix} \tag{32}$$

The determinant of the matrix on the left-hand side of eq.(32) is found as an expansion about the first row as shown below:

$$\left\| \begin{bmatrix} \frac{8}{3} & -2 & -1 & \frac{1}{3} \\ -\frac{7}{3} & 3 & \frac{7}{3} & -\frac{4}{3} \\ -2 & 4 & \frac{1}{3} & -\frac{2}{3} \\ 1 & -2 & -1 & 2 \end{bmatrix} \right\|$$

$$= \frac{8}{3} \cdot \left| \begin{bmatrix} 3 & \frac{7}{3} & -\frac{4}{3} \\ 4 & \frac{1}{3} & -\frac{2}{3} \\ -2 & -1 & 2 \end{bmatrix} \right| + 2 \cdot \left| \begin{bmatrix} -\frac{7}{3} & \frac{7}{3} & -\frac{4}{3} \\ -2 & \frac{1}{3} & -\frac{2}{3} \\ 1 & -1 & 2 \end{bmatrix} \right|$$

$$- 1 \cdot \left| \begin{bmatrix} -\frac{7}{3} & 3 & -\frac{4}{3} \\ -2 & 4 & -\frac{2}{3} \\ 1 & -2 & 2 \end{bmatrix} \right| - \frac{1}{3} \cdot \left| \begin{bmatrix} -\frac{7}{3} & 3 & \frac{7}{3} \\ -2 & 4 & \frac{1}{3} \\ 1 & -2 & -1 \end{bmatrix} \right| \qquad (33)$$

$$= \frac{8}{3} \cdot \frac{-100}{9} + 2 \cdot \frac{50}{9} - 1 \cdot \frac{-50}{9} - \frac{1}{3} \cdot \frac{25}{9} = \frac{-125}{9}$$

The inverse of the matrix is now

$$\begin{bmatrix} \frac{8}{3} & -2 & -1 & \frac{1}{3} \\ -\frac{7}{3} & 3 & \frac{7}{3} & -\frac{4}{3} \\ -2 & 4 & \frac{1}{3} & -\frac{2}{3} \\ 1 & -2 & -1 & 2 \end{bmatrix}^{-1} = -\frac{9}{125} \cdot \text{adj} \begin{bmatrix} \frac{8}{3} & -2 & -1 & \frac{1}{3} \\ -\frac{7}{3} & 3 & \frac{7}{3} & -\frac{4}{3} \\ -2 & 4 & \frac{1}{3} & -\frac{2}{3} \\ 1 & -2 & -1 & 2 \end{bmatrix}$$

$$= \frac{1}{5} \begin{bmatrix} 4 & 2 & 1 & 1 \\ 2 & 1 & 2 & 1 \\ 2 & 4 & -1 & 2 \\ 1 & 2 & 1 & 4 \end{bmatrix} \qquad (34)$$

Hence the solution to the original set of equations is given by,

$$
\begin{bmatrix} x_1 \\ x_2 \\ x_3 \\ x_4 \end{bmatrix} = \frac{1}{5} \begin{bmatrix} 4 & 2 & 1 & 1 \\ 2 & 1 & 2 & 4 \\ 2 & 4 & -1 & 2 \\ 2 & 2 & 1 & 4 \end{bmatrix} \begin{bmatrix} 2 \\ 7 \\ 6 \\ -8 \end{bmatrix} = \begin{bmatrix} 4 \\ 3 \\ 2 \\ -2 \end{bmatrix} \tag{35}
$$

The inversion technique described above, as mentioned earlier, is only intended as a very simple example. Conventional techniques of inversion like the Gauss-Jordan, Gauss-Seidel and Pivotal condensation methods are not only inefficient but also give erroneous results in the case of ill-conditioned matrices where the eigenvalue spread is large. In Section 32 we will describe more efficient methods like the Singular Value Decomposition.

Cramer's Rule

Eq.(35) presents the solution for the entire **x** vector. However, if the solution for any component x_i is desired, we can use Cramer's rule. From eq.(30) we have,

$$
x_i = \sum_{j=1}^{n} \left(\frac{\Delta_{ji}}{\det \mathbf{A}} \right) y_j \tag{36}
$$

If in the numerator of eq.(36) y_j is replaced with x_{ji}, then we have det **A** given as a cofactor expansion about the ith column as given below.

$$
\det \mathbf{X} = \sum_{j=1}^{n} (\Delta_{ji})\, x_{ji} \tag{37}
$$

However, we have y_j instead of x_{ji}. In other words, the numerator of eq.(36) is also a determinant except that the ith column is replaced by the **y** vector. Thus eq.(36) is a ratio of two determinants, the denominator being the determinant of the original **A** matrix and the numerator is the determinant of the **A** matrix with the ith column replaced by the **y** vector.

Example

We shall find the solution for x_3 in eq.(31) using Cramer's rule. Using eq.(36) we can write,

$$x_3 = \frac{\left|\begin{bmatrix} \frac{8}{3} & -2 & 2 & \frac{1}{3} \\ -\frac{7}{3} & 3 & 7 & -\frac{4}{3} \\ -2 & 4 & 6 & -\frac{2}{3} \\ 1 & -2 & -8 & 2 \end{bmatrix}\right|}{\left|\begin{bmatrix} \frac{8}{3} & -2 & -1 & \frac{1}{3} \\ -\frac{7}{3} & 3 & \frac{7}{3} & -\frac{4}{3} \\ -2 & 4 & \frac{1}{3} & -\frac{2}{3} \\ 1 & -2 & -1 & 2 \end{bmatrix}\right|} \tag{38}$$

$$= \frac{\frac{-250}{9}}{\frac{-125}{9}} = 2$$

Properties of Inverses (X, Y, Z same order and nonsingular)

1. $\mathbf{X}\,\mathbf{X}^{-1} = \mathbf{X}^{-1}\mathbf{X}$ matrix commutes with its inverse
2. $\mathbf{X}^{m}\mathbf{X}^{n} = \mathbf{X}^{m+n}$
3. $(\mathbf{XY})^{-1} = \mathbf{Y}^{-1}\mathbf{X}^{-1}$
4. $\mathbf{XY} = \mathbf{XZ} \Rightarrow \mathbf{Y} = \mathbf{Z}$
5. $(\mathbf{X}^{T})^{-1} = (\mathbf{X}^{-1})^{T}$
6. $\det \mathbf{X}^{-1} = \dfrac{1}{\det \mathbf{X}}$

Orthogonal Matrices

The matrix is **X** orthonormal if $\mathbf{X}^T\mathbf{X} = \mathbf{I}$. Since $\mathbf{X}^{-1}\mathbf{X}$ is also equal to the identity matrix we have

$$\begin{aligned} \mathbf{X}^T\mathbf{X} &= \mathbf{I} = \mathbf{X}^{-1}\mathbf{X} \\ \mathbf{X}^T &= \mathbf{X}^{-1} \end{aligned} \tag{39}$$

It immediately follows from eq.(39) that the determinant of an orthogonal matrix is equal to ± 1. Note that the converse is not true. In an orthogonal matrix **X**, individual row or column vectors are orthonormal, $\mathbf{x}_i^T\,\mathbf{x}_i = 1$. As an example, the DFT matrix is an orthogonal matrix given by

$$\mathbf{W} = \frac{1}{\sqrt{n}}\begin{bmatrix} 1 & 1 & 1 & \cdots & \cdots & 1 \\ 1 & e^{-j\frac{2\pi}{n}} & e^{-j\frac{2\pi}{n}2} & \cdots & \cdots & e^{-j\frac{2\pi}{n}(n-1)} \\ 1 & e^{-j\frac{2\pi}{n}2} & e^{-j\frac{2\pi}{n}4} & \cdots & \cdots & e^{-j\frac{2\pi}{n}2(n-1)} \\ \vdots & \vdots & \vdots & \vdots & \vdots & \\ 1 & e^{-j\frac{2\pi}{n}(n-1)} & e^{-j\frac{2\pi}{n}(n-1)2} & \cdots & \cdots & e^{-j\frac{2\pi}{n}(n-1)^2} \end{bmatrix} \tag{40}$$

It can be shown that the rows and columns of **W** are orthonormal and det **W** = 1.

Gram Matrix and Gram Determinant

One of the checks for linear independence of vectors $\{\mathbf{x}_1, \mathbf{x}_2, \ldots \mathbf{x}_m\}$ is provided by the determinant of the Gram matrix. If these vectors are arranged in an n × m matrix $\mathbf{X} = [\mathbf{x}_1, \mathbf{x}_2, \ldots \mathbf{x}_m]$, then the Gram matrix **G** is the inner product $\mathbf{X}^T\mathbf{X}$. The Gram matrix plays a key role in the singular value decomposition to be described in Section 32. The determinant of the Gram matrix det **G** is called the Gram determinant or the *Gramian*. If the Gramian is nonzero, then the m vectors $\{\mathbf{x}_1, \mathbf{x}_2, \ldots \mathbf{x}_m\}$ are linearly independent. If the Gramian is zero, then the numbers of linearly independent vectors are given by the maximum order of the matrix that gives a nonzero determinant obtained by successively deleting rows and columns of the Gramian.

Example

Let us take the example of the matrix of eq.(22) and use the Gramian to determine rank.

$$\mathbf{X} = \begin{bmatrix} 6 & 11 & 5 & 4 & 1 \\ 5 & 3 & -2 & -3 & 4 \\ -8 & -12 & -4 & 2 & -5 \\ 6 & 7 & 1 & 0 & 3 \\ 3 & 0 & -3 & -1 & 2 \end{bmatrix} \tag{22}$$

The Gramian matrix $\mathbf{G} = \mathbf{X}^T\mathbf{X}$ is given by:

$$\mathbf{G} = \mathbf{X}^T\mathbf{X}$$

$$= \begin{bmatrix} 6 & 5 & -8 & 6 & 3 \\ 11 & 3 & -12 & 7 & 0 \\ 5 & -2 & -4 & 1 & -3 \\ 4 & -3 & 2 & 0 & -1 \\ 1 & 4 & -5 & 3 & 2 \end{bmatrix} \begin{bmatrix} 6 & 11 & 5 & 4 & 1 \\ 5 & 3 & -2 & -3 & 4 \\ -8 & -12 & -4 & 2 & -5 \\ 6 & 7 & 1 & 0 & 3 \\ 3 & 0 & -3 & -1 & 2 \end{bmatrix}$$

$$
= \begin{bmatrix} 170 & 219 & 49 & -10 & 90 \\ 219 & 323 & 104 & 11 & 104 \\ 49 & 104 & 55 & 21 & 14 \\ -10 & 11 & 21 & 30 & -20 \\ 90 & 104 & 14 & -20 & 55 \end{bmatrix} \tag{41}
$$

If we take the determinant of **G** we have,

$$
|\mathbf{G}| = \left| \begin{bmatrix} 170 & 219 & 49 & -10 & 90 \\ 219 & 323 & 104 & 11 & 104 \\ 49 & 104 & 55 & 21 & 14 \\ -10 & 11 & 21 & 30 & -20 \\ 90 & 104 & 14 & -20 & 55 \end{bmatrix} \right| = 0 \tag{42}
$$

If we delete the first row and the first column of **G**, then the determinant of the resulting 4×4 matrix is,

$$
\left| \begin{bmatrix} 323 & 104 & 11 & 104 \\ 104 & 55 & 21 & 14 \\ 11 & 21 & 30 & -20 \\ 104 & 14 & -20 & 55 \end{bmatrix} \right| = 0 \tag{43}
$$

If we again delete the first row and the first column of the 4×4 matrix in eq.(43) then the determinant of the resulting 3×3 matrix is not zero as shown below

$$
\left| \begin{bmatrix} 55 & 21 & 14 \\ 21 & 30 & -20 \\ 14 & -20 & 55 \end{bmatrix} \right| = 26855 \neq 0 \tag{44}
$$

Thus, there are linearly independent vectors in the matrix **X** and hence the rank is 3, a result that has been previously derived.

31. EIGENVALUES AND EIGENVECTORS OF MATRICES

Eigenvalues

We have discussed eigen analysis in finding series solutions of differential equations in Section 6. We have similar concepts in matrices. We now investigate solutions to a linear system of equations given by,

$$\mathbf{A}\,\mathbf{x} = \lambda\,\mathbf{I}\,\mathbf{x} = \lambda\,\mathbf{x} \tag{1}$$

where $\mathbf{A}$ is an $n \times n$ matrix and $\mathbf{x}$ is an $n \times 1$ vector and λ is a scalar constant. Simplification of eq.(1) yields,

$$(\mathbf{A} - \lambda\,\mathbf{I})\,\mathbf{x} = 0 \tag{2}$$

A trivial solution to eq.(2) is $\mathbf{x} = 0$. Nontrivial solutions can exist only if

$$\det(\mathbf{A} - \lambda\,\mathbf{I}) = 0 \tag{3}$$

Expanding the determinant, the left-hand side of eq.(3) yields a polynomial of degree n in λ called the *characteristic polynomial* of the matrix **A**. Eq.(3) is called the *characteristic equation* of the matrix **A** and is of the form,

$$\lambda^{n} + a_{n-1}\lambda^{n-1} + a_{n-2}\lambda^{n-2} + \ldots + a_0\lambda^{0} = 0 \tag{4}$$

The characteristic equation has a fundamental significance in linear systems. The n roots of the characteristic equation are called the *eigenvalues* of the matrix **A** and in linear systems they are also known as the poles of the transfer function. The eigenvalues can be real, repeated or complex. From the well-known result for the roots of polynomials we also have the sum of the eigenvalues is equal to $-a_{n-1}$, and the product of the eigenvalues is equal to $(-1)^n a_0$. The product of the eigenvalues is also the determinant of the matrix **A**.

Example

Find the eigenvalues of the following matrix:

$$\mathbf{A} = \begin{bmatrix} 1 & 3 & 5 & 4 & 1 \\ 2 & -1 & -2 & -3 & 4 \\ -1 & 4 & -4 & 2 & -5 \\ 3 & 2 & 1 & 0 & 3 \\ 4 & 0 & -3 & -1 & 2 \end{bmatrix} \quad (5)$$

The characteristic equation of **A** is given by,

$$\det(\lambda\mathbf{I} - \mathbf{A}) = \begin{vmatrix} \lambda - 1 & -3 & -5 & -4 & -1 \\ -2 & \lambda + 1 & 2 & 3 & -4 \\ 1 & -4 & \lambda + 4 & -2 & 5 \\ -3 & -2 & -1 & \lambda & -3 \\ -4 & 0 & 3 & 1 & \lambda - 2 \end{vmatrix} \quad (6)$$

$$= \lambda^5 + 2\lambda^4 - 26\lambda^3 - 45\lambda^2 - 564\lambda - 408 = 0$$

The solutions of the fifth order polynomial of eq.(6) are the eigenvalues of **A** and they are:

$\lambda_1 = 6.143$, $\lambda_2 = -0.385 - j\,3.640$,

$\lambda_3 = -0.385 + j\,3.640$, $\lambda_4 = -6.626$, $\lambda_5 = -0.748$

Here two of the eigenvalues are complex conjugates of each other. The product of the eigenvalues is the determinant given by $(-1)^5.(-408) = 408$.

We can also find the eigenvalues of a square matrix that is singular. In this case the number of nonzero eigenvalues represent the rank of the matrix with the number of zero eigenvalues representing the degree of singularity.

Cayley Hamilton Theorem

Cayley Hamilton theorem states that every square matrix satisfies its own characteristic equation. In eq.(4) if we substitute the matrix **A** for λ, we obtain the Cayley Hamilton Equation.

$$\mathbf{A}^{n} + a_{n-1}\mathbf{A}^{n-1} + a_{n-2}\mathbf{A}^{n-2} + \ldots + a_0\mathbf{A}^{0} = 0 \qquad (7)$$

Example

For the matrix $\mathbf{A} = \begin{bmatrix} 0 & 1 \\ -2 & -3 \end{bmatrix}$

the characteristic equation is $\lambda^2+3\lambda+2 = 0$. According to Cayley Hamilton Theorem we should have,

$$\begin{bmatrix} 0 & 1 \\ -2 & -3 \end{bmatrix}^2 + 3\begin{bmatrix} 0 & 1 \\ -2 & -3 \end{bmatrix}^1 + 2\begin{bmatrix} 0 & 1 \\ -2 & -3 \end{bmatrix}^0$$

$$= \begin{bmatrix} -2 & -3 \\ 6 & 7 \end{bmatrix} + \begin{bmatrix} 0 & 3 \\ -6 & -9 \end{bmatrix} + \begin{bmatrix} 2 & 0 \\ 0 & 2 \end{bmatrix}$$

$$= \begin{bmatrix} 0 & 0 \\ 0 & 0 \end{bmatrix}$$

showing the validity of the theorem.

Definiteness of Matrices

A matrix **A** is positive definite if all its eigenvalues are either positive or have positive real parts. For example, the matrix **A** whose eigenvalues are given by eq.(7) is not positive definite. It is non-definite. Similarly, matrices having eigenvalues that are negative or have negative real parts are known as negative definite. If some of the eigenvalues are zero, then the matrix is called positive semi-definite or negative semi-definite correspondingly.

Since the determinants are $(-1)^n$ times the product of eigenvalues, definiteness can also be defined by the sign of the determinants of all principle minors of a matrix. Principle minors are those minors obtained by successively deleting the rows and columns of a matrix starting from the first row and first column. If all the determinants of principle minors are positive, then the matrix is positive definite; if they are negative, then it is negative definite. If they are neither positive nor negative, then it is nondefinite.

Example

We shall find the eigenvalues of the following matrix and check its definiteness:

$$\mathbf{A} = \begin{bmatrix} 6 & 11 & 5 & 4 & 1 \\ 5 & 3 & -2 & -3 & 4 \\ -8 & -12 & -4 & 2 & -5 \\ 6 & 7 & 1 & 0 & 3 \\ 3 & 0 & -3 & -1 & 2 \end{bmatrix} \quad (8)$$

As in the last example setting det $(\lambda\mathbf{I} - \mathbf{A}) = 0$ gives,

$$\det(\lambda\mathbf{I} - \mathbf{A}) = \left| \begin{bmatrix} \lambda - 6 & -11 & -5 & -4 & -1 \\ -5 & \lambda - 3 & 2 & 3 & -4 \\ 8 & 12 & \lambda + 4 & -2 & 5 \\ -6 & -7 & -1 & \lambda & -3 \\ -3 & 0 & 3 & 1 & \lambda - 2 \end{bmatrix} \right| \quad (9)$$

$$= \lambda^2 (\lambda^3 - 7\lambda^2 - 67\lambda + 1) = 0$$

Thus the rank of the matrix **A** is 3 since there are two zero eigenvalues. The eigenvalues are given by,

$\lambda_1 = 12.398$, $\lambda_2 = -5.413$, $\lambda_3 = 0.0149$

$\lambda_4 = 0$, $\lambda_5 = 0$.

Here the matrix **A** is non-definite since it has positive, negative and zero eigenvalues.

Eigenvectors

Eigenvectors are solutions to the eq.(2) for any given eigenvalue. We shall assume that all eigenvalues are distinct and there are no repeated eigenvalues. If we substitute any eigenvalue λ_i in det $(\lambda\mathbf{I} - \mathbf{A}) = 0$, then by the very definition we have det $(\lambda_i\mathbf{I} - \mathbf{A}) = 0$. Hence, solutions to eq.(2) are possible and these solutions are the eigenvectors ψ_i. However, these solutions are unique only up to a multiplicative constant. To make the solutions unique we can normalize these eigenvectors and the normalized eigenvectors are denoted by:

$$\phi_i = \frac{\psi_i}{|\psi_i|}$$

Example

Find the eigenvectors of the nonsingular matrix of eq.(5) repeated below.

$$\mathbf{A} = \begin{bmatrix} 1 & 3 & 5 & 4 & 1 \\ 2 & -1 & -2 & -3 & 4 \\ -1 & 4 & -4 & 2 & -5 \\ 3 & 2 & 1 & 0 & 3 \\ 4 & 0 & -3 & -1 & 2 \end{bmatrix} \qquad (10)$$

We shall solve for the eigenvector corresponding to the eigenvalue $\lambda_1 = 6.143$. Substituting λ_1 in eq.(2) we have,

$$(\lambda\mathbf{I} - \mathbf{A})\,\psi_1$$
$$= \begin{bmatrix} 5.143 & -3 & -5 & -4 & -1 \\ -2 & 7.143 & 2 & 3 & -4 \\ 1 & -4 & 10.143 & -2 & 5 \\ -3 & -2 & -1 & 6.143 & -3 \\ -4 & 0 & 3 & 1 & 4.143 \end{bmatrix} \begin{bmatrix} \psi_{11} \\ \psi_{12} \\ \psi_{13} \\ \psi_{14} \\ \psi_{15} \end{bmatrix} = 0 \qquad (11)$$

As mentioned before, eq.(11) can only be solved for 4 of the unknowns in terms of the fifth. Let us assume that $\psi_{11} = 1$ and solve for ψ_{12}, ψ_{13}, ψ_{14}, ψ_{15} in terms of ψ_{11}. Rearranging eq.(11) we have

$$\begin{bmatrix} 7.143 & 2 & 3 & -4 \\ -4 & 10.143 & -2 & 5 \\ -2 & -1 & 6.143 & -3 \\ 0 & 3 & 1 & 4.143 \end{bmatrix} \begin{bmatrix} \psi_{12} \\ \psi_{13} \\ \psi_{14} \\ \psi_{15} \end{bmatrix} = \begin{bmatrix} -2 \\ 1 \\ -3 \\ -4 \end{bmatrix} \qquad (12)$$

$$\begin{bmatrix} \psi_{12} \\ \psi_{13} \\ \psi_{14} \\ \psi_{15} \end{bmatrix} = \begin{bmatrix} -0.382 \\ 0.168 \\ -0.998 \\ -0.846 \end{bmatrix} \tag{13}$$

Thus, the unnormalized eigenvector corresponding to the eigenvalue $\lambda_1 = 6.143$ is given by

$$\psi_1 = \begin{bmatrix} \psi_{11} \\ \psi_{12} \\ \psi_{13} \\ \psi_{14} \\ \psi_{15} \end{bmatrix} = \begin{bmatrix} 1 \\ -0.3820 \\ 0.1682 \\ -0.9982 \\ -0.8462 \end{bmatrix} \tag{14}$$

Normalization of eq.(14) yields the eigenvector ϕ_1.

$$\phi_1 = \frac{1}{|\psi_1|} \begin{bmatrix} \psi_{11} \\ \psi_{12} \\ \psi_{13} \\ \psi_{14} \\ \psi_{15} \end{bmatrix} = \frac{1}{1.699} \begin{bmatrix} 1 \\ -0.382 \\ 0.168 \\ -0.998 \\ -0.846 \end{bmatrix} = \begin{bmatrix} 0.589 \\ -0.225 \\ 0.099 \\ -0.588 \\ -0.498 \end{bmatrix} \tag{15}$$

In a similar manner the other eigenvectors corresponding to $\lambda_2 = -0.385 - j\,3.6405$, $\lambda_3 = -0.385 + j\,3.640$, $\lambda_4 = -6.626$ and $\lambda_5 = -0.748$ are shown below.

$$\phi_2 = \begin{bmatrix} 0.2270 - j\,0.5540 \\ -0.3480 + j\,0.3900 \\ -0.2200 - j\,0.3070 \\ -0.0562 - j\,0.0404 \\ 0.0365 + j\,0.4690 \end{bmatrix} \quad \phi_3 = \begin{bmatrix} 0.2270 + j\,0.5540 \\ -0.3480 - j\,0.3900 \\ -0.2200 + j\,0.3070 \\ -0.0562 + j\,0.0404 \\ 0.0365 - j\,0.4690 \end{bmatrix}$$

$$\phi_4 = \begin{bmatrix} -0.4940 \\ 0.0415 \\ 0.7230 \\ -0.1100 \\ 0.4680 \end{bmatrix} \qquad \phi_5 = \begin{bmatrix} -0.4710 \\ -0.0949 \\ -0.3700 \\ 0.6140 \\ 0.5053 \end{bmatrix} \qquad (16)$$

Note that we have a pair of complex conjugate eigenvectors corresponding to the complex conjugate eigenvalues.

Modal Matrix

A matrix $\mathbf{\Phi}$ consisting of the linearly independent eigenvectors of a matrix **A** is called the *Modal matrix* of **A**. In the previous example the modal matrix $\mathbf{\Phi}$ is:

$$\mathbf{\Phi} = \begin{bmatrix} 0.589 & 0.227 - j\,0.554 & 0.227 + j\,0.554 & -0.494 & -0.471 \\ -0.225 & -0.348 + j\,0.390 & -0.348 - j\,0.390 & 0.0415 & -0.0949 \\ 0.0991 & -0.220 - j\,0.307 & -0.220 + j\,0.307 & 0.723 & -0.370 \\ -0.588 & -0.0562 - j\,0.0404 & -0.0562 + j\,0.0404 & -0.110 & 0.614 \\ -0.498 & 0.0365 + j\,0.469 & 0.0365 - j\,0.469 & 0.468 & 0.505 \\ \phi_1 & \phi_2 & \phi_3 & \phi_4 & \phi_5 \end{bmatrix} \qquad (17)$$

The determinant of $\mathbf{\Phi} = j\,0.403$ and it is imaginary.

Diagonalization of Matrices

An n × n square matrix **A** is diagonalizable if there exists another nonsingular n × n transformation matrix **T** such that $\mathbf{T}^{-1}\mathbf{AT} = \mathbf{D}$, where **D** is a diagonal matrix. The conditions of diagonalizability reduce to the existence of n linearly independent eigenvectors of **A** or the modal matrix Φ of **A** is invertible in which case the jth diagonal element of **D** is the jth eigenvalue of **A**. In this case, the modal matrix Φ is the desired matrix **T**. However, the modal matrix Φ is not the only matrix that will diagonalize **A**. There are other matrices that may diagonalize **A**, but only the diagonal elements obtained from the modal transformation matrix are the eigenvalues of **A**.

As a consequence of the above discussion we can write the matrix equivalent of the eigen equation (1) as follows.

$\mathbf{A}\Phi = \Phi\,\Lambda$ or in vector form,

$$\mathbf{A}\boldsymbol{\phi}_{\mathbf{k}} = \boldsymbol{\phi}_{\mathbf{k}}\,\lambda_k \tag{18}$$

Example

Diagonalize the following matrix:

$$\mathbf{A} = \begin{bmatrix} 1 & 0 & 1 & -1 \\ 1 & -1 & 0 & -1 \\ 1 & 1 & -1 & 1 \\ 0 & -1 & 1 & 0 \end{bmatrix} \tag{19}$$

The eigenvalues of the matrix are obtained by solving the determinant equation det $|\lambda\mathbf{I} - \mathbf{A}| = 0$. Or,

$$\det(\lambda\mathbf{I} - \mathbf{A}) = \left\| \begin{bmatrix} \lambda - 1 & 0 & -1 & 1 \\ -1 & \lambda + 1 & 0 & 1 \\ -1 & -1 & \lambda + 1 & -1 \\ 0 & 1 & -1 & \lambda \end{bmatrix} \right\|$$

$$= \lambda^4 + \lambda^3 - 4\,\lambda^2 - 2\,\lambda + 4 = 0$$

The characteristic equation can be factored as $(\lambda-\sqrt{2})(\lambda+\sqrt{2})(\lambda-1)(\lambda+2) = 0$ yielding the eigenvalues as: $\lambda_1 = \sqrt{2}$, $\lambda_2 = -\sqrt{2}$, $\lambda_3 = 1$, $\lambda_4 = -2$. We will now find the eigenvectors corresponding to these eigenvalues. Let us go through the steps of finding the eigenvector for the eigenvalue $\lambda_1 = \sqrt{2}$. Substituting for λ_1 in eq.(18), we obtain,

$$\begin{bmatrix} \sqrt{2}-1 & 0 & -1 & 1 \\ -1 & \sqrt{2}+1 & 0 & 1 \\ -1 & -1 & \sqrt{2}+1 & -1 \\ 0 & 1 & -1 & \sqrt{2} \end{bmatrix} \begin{bmatrix} \psi_{11} \\ \psi_{12} \\ \psi_{13} \\ \psi_{14} \end{bmatrix} = 0 \qquad (20)$$

Assuming $\psi_{11} = 1$ we solve for $\psi_{12}, \psi_{13}, \psi_{14}$ as,

$$\psi_1 = \begin{bmatrix} 1 & \dfrac{\sqrt{2}}{3+\sqrt{2}} & \dfrac{2\sqrt{2}}{3+\sqrt{2}} & \dfrac{1}{3+\sqrt{2}} \end{bmatrix}^T$$

and normalizing ψ_1 we obtain ϕ_1 as,

$$\phi_1 = \frac{\psi_1}{|\psi_1|} = \begin{bmatrix} \dfrac{\sqrt{2}\left(3+\sqrt{2}\right)}{2\left(11+3\sqrt{2}\right)} \\ \dfrac{1}{\left(11+3\sqrt{2}\right)} \\ \dfrac{2}{\left(11+3\sqrt{2}\right)} \\ \dfrac{\sqrt{2}}{\left(11+3\sqrt{2}\right)} \end{bmatrix} = \begin{bmatrix} 0.799 \\ 0.256 \\ 0.512 \\ 0.181 \end{bmatrix} \qquad (21)$$

In a similar manner we can find the other eigenvectors ϕ_2, ϕ_3, ϕ_4.

We can now find the modal matrix $\Phi = [\phi_1, \phi_2, \phi_3, \phi_4]$

$$\Phi = \begin{bmatrix} \phi_{11} & \phi_{21} & \phi_{31} & \phi_{41} \\ \phi_{12} & \phi_{22} & \phi_{32} & \phi_{42} \\ \phi_{13} & \phi_{23} & \phi_{33} & \phi_{43} \\ \phi_{14} & \phi_{24} & \phi_{34} & \phi_{44} \end{bmatrix} = \begin{bmatrix} 0.800 & 0.431 & -0.577 & -0.408 \\ 0.256 & -0.385 & 0 & 0 \\ 0.512 & -0.769 & -0.577 & 0.817 \\ 0.181 & 0.272 & -0.577 & -0.408 \end{bmatrix} \quad (22)$$

and diagonalization using **Φ** results in,

$$\Phi^{-1}\mathbf{A}\Phi = \begin{bmatrix} \sqrt{2} & 0 & 0 & 0 \\ 0 & -\sqrt{2} & 0 & 0 \\ 0 & 0 & 1 & 0 \\ 0 & 0 & 0 & -2 \end{bmatrix} \quad (23)$$

Symmetric Matrices

Real symmetric matrices are those matrices that satisfy the condition,

$$\mathbf{A}^T = \mathbf{A} \quad (24)$$

Complex symmetric matrices satisfy the Hermitian symmetry condition,

$$\mathbf{A}^H = (\mathbf{A}^*)^T = \mathbf{A} \quad (25)$$

the symbol **H** standing for the conjugate transpose.

Many applications require symmetric matrices. As an example, the covariance matrices in probability are all real symmetric matrices. Some of the properties of symmetric matrices are used in simplifying algebraic manipulations. The more important properties of symmetric matrices are:

1. All eigenvalues are real.
2. The eigenvectors corresponding to distinct eigenvalues are orthogonal. Hence the eigenvectors form

the basis vectors of an n-dimensional Euclidean coordinate system.

3. Since modal matrix $\mathbf{\Phi}$ is orthogonal and the diagonalization of a symmetric matrix similar to eq.(23) can be accomplished by the following transformation, $\mathbf{\Phi}^T\mathbf{A}\mathbf{\Phi} = \Lambda$.

Example

Let us find the eigenvectors and eigenvalues of the symmetric matrix given below.

$$\mathbf{A} = \begin{bmatrix} 1 & -2 & 1 & -1 \\ -2 & 1 & 1 & -1 \\ 1 & 1 & 2 & -2 \\ -1 & -1 & -2 & 0 \end{bmatrix} \tag{26}$$

The characteristic equation is det $[\lambda\mathbf{I} - \mathbf{A}] = 0$. Or,

$$\det\left(\lambda I - \mathbf{A}\right) = \left|\begin{bmatrix} \lambda - 1 & 2 & -1 & 1 \\ 2 & \lambda - 1 & -1 & 1 \\ -1 & -1 & \lambda - 2 & 2 \\ 1 & 1 & 2 & \lambda \end{bmatrix}\right| \tag{27}$$
$$= \lambda^4 - 4\,\lambda^3 - 7\,\lambda^2 + 22\,\lambda + 24$$

The eigenvalues are:

$\lambda_1 = -2,\ \lambda_2 = 3,\ \lambda_3 = 4,\ \lambda_4 = -1.$

We note in eq.(27) that the sum of the eigenvalues is 4, that is (–1) times the coefficient of λ^3, and the product of the eigenvalues is 24, that is $(-1)^4$ times the constant in the characteristic equation (27). The four unnormalized eigenvectors are obtained as outlined in the previous examples and they are,

$$\psi_1 = \begin{bmatrix} 1 \\ 1 \\ 0 \\ 1 \end{bmatrix} \quad \psi_2 = \begin{bmatrix} -1 \\ 1 \\ 0 \\ 0 \end{bmatrix} \quad \psi_3 = \begin{bmatrix} 1 \\ 1 \\ 3 \\ -2 \end{bmatrix} \quad \psi_4 = \begin{bmatrix} -1 \\ -1 \\ 2 \\ 2 \end{bmatrix} \tag{28}$$

for the eigenvalues $\lambda_1 = -2$, $\lambda_2 = 3$, $\lambda_3 = 4$ and $\lambda_4 = -1$ and the corresponding normalized eigenvectors are,

$$\phi_1 = \frac{1}{\sqrt{3}}\begin{bmatrix} 1 \\ 1 \\ 0 \\ 1 \end{bmatrix} \quad \phi_2 = \frac{1}{\sqrt{2}}\begin{bmatrix} -1 \\ 1 \\ 0 \\ 0 \end{bmatrix}$$

$$\phi_3 = \frac{1}{\sqrt{15}}\begin{bmatrix} 1 \\ 1 \\ 3 \\ -2 \end{bmatrix} \quad \phi_4 = \frac{1}{\sqrt{10}}\begin{bmatrix} -1 \\ -1 \\ 2 \\ 2 \end{bmatrix} \tag{29}$$

The modal matrix $\mathbf{\Phi}$ is given by,

$$\Phi = \frac{1}{\sqrt{3}}.\frac{1}{\sqrt{2}}.\frac{1}{\sqrt{15}}.\frac{1}{\sqrt{10}}.\begin{bmatrix} 1 & -1 & 1 & -1 \\ 1 & 1 & 1 & -1 \\ 0 & 0 & 3 & 2 \\ 1 & 0 & -2 & 2 \end{bmatrix} \tag{30}$$

with det $\mathbf{\Phi} = 1$ showing that $\mathbf{\Phi}$ is an orthogonal matrix.

The original matrix **A** can be diagonalized by the modal matrix as shown below.

$\mathbf{\Phi^T A \Phi}$

$$= \frac{1}{30^2}.\begin{bmatrix} 1 & 1 & 0 & 1 \\ -1 & 1 & 0 & 0 \\ 1 & 1 & 3 & -2 \\ -1 & -1 & 2 & 2 \end{bmatrix}.\begin{bmatrix} 1 & -2 & 1 & -1 \\ -2 & 1 & 1 & -1 \\ 1 & 1 & 2 & -2 \\ -1 & -1 & -2 & 0 \end{bmatrix}\begin{bmatrix} 1 & -1 & 1 & -1 \\ 1 & 1 & 1 & -1 \\ 0 & 0 & 3 & 2 \\ 1 & 0 & -2 & 2 \end{bmatrix} \tag{31}$$

$$= \begin{bmatrix} -2 & 0 & 0 & 0 \\ 0 & 3 & 0 & 0 \\ 0 & 0 & 4 & 0 \\ 0 & 0 & 0 & -1 \end{bmatrix}$$

Similarity Transformations

In the previous section we obtained the matrix Λ by transforming with the modal matrix $\mathbf{\Phi}$. The resulting diagonal matrix Λ has all the fundamental properties of the original matrix **A**. For example, these two matrices have the same determinants and the same traces. Such matrices are called similar matrices and the transformation is known as a similarity transformation. Any nonsingular matrix **A** can be transformed in to another similar matrix **B** by the following similarity transformation

$$\mathbf{T}^{-1}\mathbf{A}\mathbf{T} = \mathbf{B} \tag{32a}$$

where **T** is another nonsingular transforming matrix. If **T** is also orthogonal, then the transformation given by eq.(32a) can be written as

$$\mathbf{T}^{T}\mathbf{A}\mathbf{T} = \mathbf{B} \tag{32b}$$

Example

For the matrix **A** given by eq.(26) and for an arbitrary non-singular matrix **T** given by,

$$\mathbf{A} = \begin{bmatrix} 1 & -2 & 1 & -1 \\ -2 & 1 & 1 & -1 \\ 1 & 1 & 2 & -2 \\ -1 & -1 & -2 & 0 \end{bmatrix} \quad \mathbf{T} = \begin{bmatrix} 1 & 0 & 1 & -1 \\ 1 & -1 & 0 & -1 \\ 1 & 1 & -1 & 1 \\ 0 & -1 & 1 & 0 \end{bmatrix}$$

the transformation $\mathbf{B} = \mathbf{T}^{-1}\mathbf{A}\mathbf{T}$ can be written as,

$$\mathbf{B} = \mathbf{T}^{-1}\mathbf{A}\mathbf{T} = \begin{bmatrix} 1 & \frac{5}{2} & -\frac{5}{2} & 1 \\ 2 & 2 & 1 & 0 \\ -2 & 1 & 2 & 0 \\ -1 & -\frac{1}{2} & \frac{1}{2} & -1 \end{bmatrix} \tag{33}$$

However, if **T** is the modal matrix **Φ** given by eq.(30) then the transformed matrix **B** is the diagonal matrix Λ of eigenvalues as shown in eq.(31). In all the cases the determinants of **A**, **B** and Λ are 24 and the traces of **A**, **B** and Λ are 4.

Simultaneous Diagonalization of Two Symmetric Nonsingular Matrices

If we are given two positive definite *symmetric* matrices **A** and **B** of the same dimension n, with distinct eigenvalues it is useful in pattern recognition to diagonalize both matrices by means of the same transformation matrix **T**. The transforming matrix **T** will take **A** to an identity matrix while at the same time diagonalizing **B**. We shall now analyze the procedure that will accomplish this. The two positive definite symmetric matrices **A** and **B** with distinct eigenvalues are,

$$\mathbf{A} = \begin{bmatrix} a_{11} & a_{12} & \cdots & a_{1n} \\ a_{12} & a_{22} & \cdots & a_{2n} \\ \vdots & \vdots & \vdots & \vdots \\ a_{1n} & a_{2n} & \cdots & a_{nn} \end{bmatrix} \quad \mathbf{B} = \begin{bmatrix} b_{11} & b_{12} & \cdots & b_{1n} \\ b_{12} & b_{22} & \cdots & b_{2n} \\ \vdots & \vdots & \vdots & \vdots \\ b_{1n} & b_{2n} & \cdots & b_{nn} \end{bmatrix} \qquad (34)$$

We have seen in previous sections that the modal matrix **Φ** of **A** will transform **A** to a diagonal matrix of its eigenvalues $\{\lambda_1, \lambda_2, \lambda_3, \ldots, \lambda_n\}$ as shown below.

$$\mathbf{\Phi}^T\mathbf{A}\mathbf{\Phi} = \Lambda = \mathrm{diag}\{\lambda_1, \lambda_2, \lambda_3, \ldots, \lambda_n\} \qquad (35)$$

Since Λ is diagonal with positive values we take its positive square root and write,

$$\mathbf{\Phi}^T\mathbf{A}\mathbf{\Phi} = \sqrt{\Lambda}\ \mathbf{I}\sqrt{\Lambda},$$

$$\text{Or, } (\sqrt{\Lambda})^{-1}\mathbf{\Phi}^T\mathbf{A}\mathbf{\Phi}(\sqrt{\Lambda})^{-1} = \mathbf{I} \qquad (36)$$

We now define $\mathbf{U} = \mathbf{\Phi}(\sqrt{\Lambda})^{-1}$ and write eq.(36) as,

$$(\sqrt{\Lambda})^{-1}\mathbf{\Phi}^T\mathbf{A}\mathbf{\Phi}(\sqrt{\Lambda})^{-1} = \mathbf{U}^T\mathbf{A}\mathbf{U} = \mathbf{I} \qquad (37)$$

In eq.(37) we have not only diagonalized **A** but also reduced it to an identity matrix. This process is known as whitening the matrix **A**. Since **B** is a symmetric matrix, we can show that $\mathbf{U^TBU}$ is also a symmetric matrix as shown below:

$$\left[(\sqrt{\Lambda})^{-1}\mathbf{\Phi^T B \Phi}(\sqrt{\Lambda})^{-1}\right]^T = (\sqrt{\Lambda})^{-1}\mathbf{\Phi^T B^T \Phi}(\sqrt{\Lambda})^{-1}$$
$$= (\sqrt{\Lambda})^{-1}\mathbf{\Phi^T B \Phi}(\sqrt{\Lambda})^{-1}$$

The symmetric matrix $\mathbf{U^TBU}$ can, therefore, be diagonalized by its orthogonal modal matrix, **M**, that is, $\mathbf{M^TU^TBUM} = \mathbf{D} = \text{diag}\{d_1, d_2, \ldots d_n\}$ where $\{d_i\}$ are the distinct eigenvalues of $\mathbf{U^TBU}$. The transformation **UM** has diagonalized **B** and now we show that **UM** applied to **A** also retains the whitening transformation. Using eq.(37) we can write,

$$\mathbf{M^TU^TAUM} = \mathbf{M^T}(\sqrt{\Lambda})^{-1}\mathbf{\Phi^T A \Phi}\,(\sqrt{\Lambda})^{-1}\,\mathbf{M}$$
$$= \mathbf{M^TIM} = \mathbf{I} \qquad (38)$$

Hence the transformation,

$$\mathbf{T} = \mathbf{\Phi}\,(\sqrt{\Lambda})^{-1}\,\mathbf{M} = \mathbf{UM} \qquad (39)$$

is the required transformation that takes **A** into an identity matrix while taking **B** into a diagonal matrix. In the transformation **T** in eq.(39), $\mathbf{\Phi}$ is the modal matrix of **A**, Λ is the diagonal matrix of eigenvalues of **A**, and **M** is the modal matrix of

$$(\sqrt{\Lambda})^T\mathbf{\Phi^T B \Phi}\,\sqrt{\Lambda} = \mathbf{U^TBU}.$$

In practice, the above procedure can be simplified as follows. From eqs.(38, 39) we have

$$\mathbf{T^TAT} = \mathbf{I} : \det(\mathbf{T^TAT}) = 1 \qquad (40)$$

As a result we have $\mathbf{AT} = \mathbf{T^{-T}}$ (41)

But $\mathbf{T^TBT} = \mathbf{D}$ and from eq.(41)

$$\mathbf{BT} = \mathbf{T^{-T}D} = \mathbf{ATD} \qquad (42)$$

Eq.(42) expressed as

$$\mathbf{A}^{-1}\mathbf{B}\mathbf{T} = \mathbf{T}\mathbf{D} \tag{43}$$

is in the form of an eigen equation (18) with **T** serving as the modal matrix of the product matrix $\mathbf{A}^{-1}\mathbf{B}$, and **D** is the diagonal eigenvalue matrix of $\mathbf{A}^{-1}\mathbf{B}$.

In summary, the steps for computation of the transformation matrix **T** that diagonalizes the two n × n symmetric positive definite matrices **A** and **B** are as follows.

1. Calculate the n distinct eigenvalues $\{d_1, d_2, \ldots d_n\}$ of the matrix $\mathbf{A}^{-1}\mathbf{B}$ from $\det(\lambda\mathbf{I} - \mathbf{A}^{-1}\mathbf{B}) = 0$.
2. Calculate the eigenvectors τ_i from $(\lambda\mathbf{I} - \mathbf{A}^{-1}\mathbf{B})\tau_i = 0$.
3. Form the modal matrix $N = \{\tau_1, \tau_2, \ldots, \tau_n\}$.
4. Form the matrix $\mathbf{W} = N^T\mathbf{A}N$ (normalization with respect to **A**).
5. **W** is diagonal with positive eigenvalues. Hence rewrite 4, as $(\sqrt{\mathbf{W}})^{-1} N^T\mathbf{A}N (\sqrt{\mathbf{W}})^{-1} = \mathbf{I}$.
6. The desired transformation is $\mathbf{T} = N(\sqrt{\mathbf{W}})^{-1}$.

Example

Find the transformation matrix **T** for the symmetric matrices **A** and **B** shown on the next page. The corresponding eigenvalues and the determinants are also shown.

$$\mathbf{A} = \begin{bmatrix} 5\,2\,1\,2 \\ 2\,4\,1\,1 \\ 1\,1\,3\,2 \\ 2\,1\,2\,2 \end{bmatrix} \qquad \mathbf{B} = \begin{bmatrix} 1\,0\,0\,1 \\ 0\,2\,1\,1 \\ 0\,1\,2\,1 \\ 1\,1\,1\,2 \end{bmatrix}$$

$$\det\mathbf{A} = 24 \qquad \det\mathbf{B} = 1 \tag{44}$$

$$\lambda_{\mathbf{A}} = \begin{bmatrix} 8.28 \\ 2.53 \\ 3.00 \\ 0.19 \end{bmatrix} \qquad \lambda_{\mathbf{B}} = \begin{bmatrix} 0.14 \\ 1.75 \\ 4.11 \\ 1.00 \end{bmatrix}$$

The matrix $\mathbf{A}^{-1}\mathbf{B}$ and its eigenvalues are shown below:

$$\mathbf{A}^{-1}\mathbf{B} = \begin{bmatrix} -\frac{1}{2} & -\frac{5}{6} & -\frac{1}{12} & -\frac{7}{6} \\ 0 & \frac{2}{3} & \frac{1}{6} & \frac{1}{3} \\ -\frac{3}{2} & -\frac{5}{6} & \frac{11}{12} & -\frac{13}{6} \\ \frac{5}{2} & \frac{11}{6} & -\frac{5}{12} & \frac{25}{6} \end{bmatrix} \tag{45}$$

$$\mathbf{D} = \Lambda_{(\mathbf{A}^{-1}\mathbf{B})} = \begin{bmatrix} 3.880 & 0 & 0 & 0 \\ 0 & 0.058 & 0 & 0 \\ 0 & 0 & 0.896 & 0 \\ 0 & 0 & 0 & 0.411 \end{bmatrix}$$

The modal matrix $\mathbf{N}$ for $\mathbf{A}^{-1}\mathbf{B}$, and $\mathbf{W} = \mathrm{N}^{T}\mathbf{A}\mathrm{N}$ are now calculated.

$$\mathbf{N} = \begin{bmatrix} -0.22 & 0.76 & -0.50 & 0.31 \\ 0.06 & 0.27 & 0.61 & -0.68 \\ -0.51 & 0.16 & 0.61 & 0.65 \\ 0.83 & -0.57 & 0.12 & 0.20 \end{bmatrix} \tag{46}$$

$$\mathbf{W} = \mathbf{N}^{T}\mathbf{A}\mathbf{N} = \begin{bmatrix} 0.2 & 0 & 0 & 0 \\ 0 & 2.66 & 0 & 0 \\ 0 & 0 & 2.97 & 0 \\ 0 & 0 & 0 & 2.78 \end{bmatrix}$$

The following equations represent $(\sqrt{\mathbf{W}})$ and $(\sqrt{\mathbf{W}})^{-1}$

$$\sqrt{\mathbf{W}} = \begin{bmatrix} 0.45 & 0 & 0 & 0 \\ 0 & 1.63 & 0 & 0 \\ 0 & 0 & 1.72 & 0 \\ 0 & 0 & 0 & 1.67 \end{bmatrix} \tag{47}$$

$$\sqrt{\mathbf{W}}^{-1} = \begin{bmatrix} 2.24 & 0 & 0 & 0 \\ 0 & 0.61 & 0 & 0 \\ 0 & 0 & 0.58 & 0 \\ 0 & 0 & 0 & 0.6 \end{bmatrix}$$

and the desired transformation matrix $\mathbf{T} = N(\sqrt{\mathbf{W}})^{-1}$ and the transformations $\mathbf{T}^T\mathbf{A}\mathbf{T} = \mathbf{I}$ and $\mathbf{T}^T\mathbf{B}\mathbf{T} = \mathbf{D}$ are shown.

$$\mathbf{T} = \begin{bmatrix} -0.50 & 0.47 & -0.29 & 0.18 \\ 0.13 & 0.16 & 0.35 & -0.40 \\ -1.14 & 0.10 & 0.35 & 0.38 \\ 1.86 & -0.35 & 0.07 & 0.12 \end{bmatrix}$$

$$\mathbf{T}^T\mathbf{A}\mathbf{T} = \begin{bmatrix} 1 & 0 & 0 & 0 \\ 0 & 1 & 0 & 0 \\ 0 & 0 & 1 & 0 \\ 0 & 0 & 0 & 1 \end{bmatrix} \tag{48}$$

$$\mathbf{T}^T\mathbf{B}\mathbf{T} = \begin{bmatrix} 3.880 & 0 & 0 & 0 \\ 0 & 0.058 & 0 & 0 \\ 0 & 0 & 0.896 & 0 \\ 0 & 0 & 0 & 0.411 \end{bmatrix}$$

Canonic Forms

We have already seen that similar matrices have certain invariant characteristics like the determinant, the trace, the eigenvalues and the characteristic equation. By proper similarity transformations, matrices can be expressed in various canonic forms that are suitable for mathematical manipulations. The diagonal matrix consisting of distinct eigenvalues of any matrix **A** is called the Jordan canonic form where all superdiagonal terms are zero. However, we have also seen that any $n \times n$ matrix **A** is diagonalizable if and only if it has n linearly independent eigenvectors. If **A** has n distinct eigenvalues, then there are n linearly independent eigenvectors guaranteeing diagonalizability and the existence of a diagonal Jordan canonic form. This may not be true if there are repeated eigenvalues. We shall explain this clearly by assuming that the $k \times k$ matrix **B** has a single eigenvalue a repeated k times. Therefore, the characteristic equation of **B** is $(\lambda - a)^k = 0$. If **B** has all k linearly independent eigenvectors, then the Jordan canonic form is a k-dimensional diagonal matrix $\mathbf{J}_1$ of the single eigenvalue a with all superdiagonal terms being zero. However, if there are only i linearly independent eigenvectors with $i < k$, then **B** can not be completely diagonalized. In addition to the i diagonal elements containing a, we will also have $j = k-i$ ones in the first superdiagonal as shown in the matrix $\mathbf{J}_2$. One form of the Jordan canonic form for repeated eigenvalues a is shown in the next page. In the matrix $\mathbf{J}_1$ the repeated eigenvalues a have independent eigenvectors and are in the diagonal form. In the matrix $\mathbf{J}_2$ only i eigenvectors are independent, and among the other $j = k-i$ eigenvectors, $j-1$ are dependent. Hence, we have $j-1$ first superdiagonal ones in the $j \times j$ sub block and the $i \times i$ sub block is diagonal.

$$
\mathbf{J}_1 = \begin{bmatrix}
\begin{bmatrix}
a & 0 & \dots & 0 & 0 \\
0 & a & \dots & 0 & 0 \\
\vdots & i & \times & i & \vdots \\
0 & 0 & \dots & a & 0 \\
0 & 0 & \dots & 0 & a
\end{bmatrix} &
\begin{bmatrix}
0 & 0 & \dots & 0 & 0 \\
0 & 0 & \dots & 0 & 0 \\
\vdots & i & \times & j & \vdots \\
0 & 0 & \dots & 0 & 0 \\
0 & 0 & \dots & 0 & 0
\end{bmatrix} \\
\begin{bmatrix}
0 & 0 & \dots & 0 & 0 \\
0 & 0 & \dots & 0 & 0 \\
\vdots & j & \times & i & \vdots \\
0 & 0 & \dots & 0 & 0 \\
0 & 0 & \dots & 0 & 0
\end{bmatrix} &
\begin{bmatrix}
a & 0 & \dots & 0 & 0 \\
0 & a & \dots & 0 & 0 \\
\vdots & j & \times & j & \vdots \\
0 & 0 & \dots & a & 0 \\
0 & 0 & \dots & 0 & a
\end{bmatrix}
\end{bmatrix}
$$

$k \times k$ square matrix

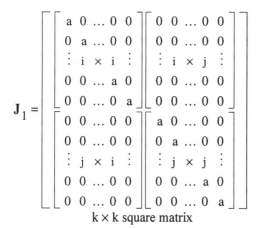

$$
\mathbf{J}_2 = \begin{bmatrix} \begin{bmatrix} a & 0 & \dots & 0 & 0 \\ 0 & a & \dots & 0 & 0 \\ \vdots & i & \times & i & \vdots \\ 0 & 0 & \dots & a & 0 \\ 0 & 0 & \dots & 0 & a \end{bmatrix} & \begin{bmatrix} 0 & 0 & \dots & 0 & 0 \\ 0 & 0 & \dots & 0 & 0 \\ \vdots & i & \times & j & \vdots \\ 0 & 0 & \dots & 0 & 0 \\ 0 & 0 & \dots & 0 & 0 \end{bmatrix} \\ \begin{bmatrix} 0 & 0 & \dots & 0 & 0 \\ 0 & 0 & \dots & 0 & 0 \\ \vdots & j & \times & i & \vdots \\ 0 & 0 & \dots & 0 & 0 \\ 0 & 0 & \dots & 0 & 0 \end{bmatrix} & \begin{bmatrix} a & 1 & \dots & 0 & 0 \\ 0 & a & \dots & 0 & 0 \\ \vdots & j & \times & j & \vdots \\ 0 & 0 & \dots & a & 1 \\ 0 & 0 & \dots & 0 & a \end{bmatrix} \end{bmatrix} \quad (49)
$$

k × k square matrix

The second canonic form is the companion form sometimes called the Frobenius canonic form. It is very useful in the state space formulation of control problems. One of the forms is shown below. The last row in the companion form of a square matrix **A** contains the negative of the coefficients of the characteristic equation with all other elements 0 except the first superdiagonal that contains 1. Thus, if the characteristic polynomial is $\lambda^n + a_{n-1}\lambda^{n-1} + a_{n-2}\lambda^{n-2} + \dots a_1\lambda + a_0$, then one form of the companion matrix is shown below.

$$
\mathbf{F} = \begin{bmatrix} 0 & 1 & \cdots & 0 & 0 \\ 0 & 0 & \cdots & 0 & 0 \\ \vdots & \vdots & \vdots & \vdots & \vdots \\ 0 & 0 & \cdots & 1 & 0 \\ 0 & 0 & \cdots & 0 & 1 \\ -a_0 & -a_1 & \cdots & -a_{n-2} & -a_{n-1} \end{bmatrix} \quad (50)
$$

If the eigenvalues are distinct, then the companion matrix **F** can be transformed into a diagonal matrix by the transformation $\mathbf{T}^{-1}\mathbf{F}\mathbf{T} = \Lambda$ where **T** is the Vandermonde matrix given by,

$$\mathbf{T} = \begin{bmatrix} 1 & 1 & \cdots & 1 & \cdots & 1 \\ \lambda_1^2 & \lambda_2^2 & \cdots & \lambda_i^2 & \cdots & \lambda_n^2 \\ \vdots & \vdots & \vdots & \vdots & \vdots & \vdots \\ \lambda_1^{i-1} & \lambda_2^{i-1} & \cdots & \lambda_i^{i-1} & \cdots & \lambda_n^{i-1} \\ \vdots & \vdots & \vdots & \vdots & \vdots & \vdots \\ \lambda_1^{n-1} & \lambda_2^{n-1} & \cdots & \lambda_i^{n-1} & \cdots & \lambda_n^{n-1} \end{bmatrix} \tag{51}$$

Example

Corresponding to the characteristic polynomial $\lambda^4 + 10\lambda^3 + 35\lambda^2 + 50\lambda + 24$ the companion matrix is given by,

$$\mathbf{F} = \begin{bmatrix} 0 & 1 & 0 & 0 \\ 0 & 0 & 1 & 0 \\ 0 & 0 & 0 & 1 \\ -24 & -50 & -35 & -10 \end{bmatrix} \tag{52}$$

The eigenvalues are:

$\lambda_1 = -4,\ \lambda_2 = -3,\ \lambda_3 = -2,\ \lambda_4 = -1.$

We will show that the Vandermonde matrix **T** and its inverse $\mathbf{T}^{-1}$ given by eqs.(53) diagonalize **F**.

$$\mathbf{T} = \begin{bmatrix} 1 & 1 & 1 & 1 \\ -4 & -3 & -2 & -1 \\ 16 & 9 & 4 & 1 \\ -64 & -27 & -8 & -1 \end{bmatrix} \quad \mathbf{T}^{-1} = \begin{bmatrix} -1 & -\frac{11}{6} & -1 & -\frac{1}{6} \\ 4 & 7 & \frac{7}{2} & \frac{1}{2} \\ -6 & -\frac{19}{2} & -4 & -\frac{1}{2} \\ 4 & \frac{13}{3} & \frac{3}{2} & \frac{1}{6} \end{bmatrix} \tag{53}$$

$$
\mathbf{T}^{-1}\mathbf{FT} = \begin{bmatrix} -1 & -\frac{11}{6} & -1 & -\frac{1}{6} \\ 4 & 7 & \frac{7}{2} & \frac{1}{2} \\ -6 & -\frac{19}{2} & -4 & -\frac{1}{2} \\ 4 & \frac{13}{3} & \frac{3}{2} & \frac{1}{6} \end{bmatrix} \begin{bmatrix} 0 & 1 & 0 & 0 \\ 0 & 0 & 1 & 0 \\ 0 & 0 & 0 & 1 \\ -24 & -50 & -35 & -10 \end{bmatrix}
$$

$$
\times \begin{bmatrix} 1 & 1 & 1 & 1 \\ -4 & -3 & -2 & -1 \\ 16 & 9 & 4 & 1 \\ -64 & -27 & -8 & -1 \end{bmatrix} = \begin{bmatrix} -4 & 0 & 0 & 0 \\ 0 & -3 & 0 & 0 \\ 0 & 0 & -2 & 0 \\ 0 & 0 & 0 & -1 \end{bmatrix} \tag{54}
$$

Block Matrices

In many applications we will encounter matrices whose elements are also matrices. Here, we have to be extra careful about the compatibility of individual blocks. In an n × n matrix **X** we can formulate it as a block matrix consisting of individual blocks **A, B, C, D**.

$$
\mathbf{X} = \begin{bmatrix} \begin{bmatrix} x_{11} & x_{12} & \cdots & x_{1m} \\ x_{21} & x_{22} & \cdots & x_{2m} \\ \vdots & & \mathbf{A} = \text{mxm} & \vdots \\ x_{m1} & x_{m2} & \cdots & x_{mm} \end{bmatrix} \\ \begin{bmatrix} x_{m+11} & x_{m+12} & \cdots & x_{m+1m} \\ x_{m+21} & x_{m+22} & \cdots & x_{m+2m} \\ \vdots & & \mathbf{C} = (n-m)\text{xm} & \vdots \\ x_{n1} & x_{n2} & \cdots & x_{nm} \end{bmatrix} \end{bmatrix}
$$

$$\left[\begin{array}{c} \begin{bmatrix} x_{1m+1} & x_{1m+2} & \cdots & x_{1n} \\ x_{2m+1} & x_{2m+2} & \cdots & x_{2n} \\ \vdots & & \mathbf{B} = \ mx(n-m) & \vdots \\ x_{mm+1} & x_{mm+2} & \cdots & x_{mn} \end{bmatrix} \\ \begin{bmatrix} x_{m+1m+1} & x_{m+1m+2} & \cdots & x_{m+1n} \\ x_{m+2m+1} & x_{m+2m+2} & \cdots & x_{m+2n} \\ \vdots & & \mathbf{D} = (n-m) \ x \ (n-m) & \vdots \\ x_{nm+1} & x_{nm+2} & \cdots & x_{nn} \end{bmatrix} \end{array}\right]$$

$$= \begin{bmatrix} \underset{mxm}{\mathbf{A}} & \underset{mx(n-m)}{\mathbf{B}} \\ \underset{(n-m)xm}{\mathbf{C}} & \underset{(n-m)x(n-m)}{\mathbf{D}} \end{bmatrix} \qquad (55)$$

Clearly, the compatibility conditions as shown in eq.(55) have to be satisfied.

Properties of block matrices

1. Multiplication

The compatibility of multiplying block matrices should be checked in the following example.

$$\begin{bmatrix} \underset{n \times m}{\mathbf{A}} & \underset{n \times p}{\mathbf{B}} \\ \underset{n \times m}{\mathbf{C}} & \underset{n \times p}{\mathbf{D}} \end{bmatrix} \begin{bmatrix} \underset{m \times r}{\mathbf{E}} & \underset{m \times q}{\mathbf{F}} \\ \underset{p \times r}{\mathbf{G}} & \underset{p \times q}{\mathbf{H}} \end{bmatrix} = \begin{bmatrix} \underset{n \times r}{\mathbf{AE} + \mathbf{BG}} & \underset{n \times q}{\mathbf{AF} + \mathbf{BH}} \\ \underset{n \times r}{\mathbf{CE} + \mathbf{DG}} & \underset{n \times q}{\mathbf{CF} + \mathbf{DH}} \end{bmatrix} \qquad (56)$$

2. Determinant

If **A** and **D** are square (not necessarily nonsingular), then

$$\det \begin{bmatrix} \mathbf{A} & 0 \\ \mathbf{C} & \mathbf{D} \end{bmatrix} = \det \mathbf{A} \ \det \mathbf{D} \qquad (57)$$

3. If **A** is nonsingular and **D** is square, then

$$\det\begin{bmatrix} \mathbf{A} & \mathbf{B} \\ \mathbf{C} & \mathbf{D} \end{bmatrix} = \det \mathbf{A} \det\left(\mathbf{D} - \mathbf{C}\mathbf{A}^{-1}\mathbf{B}\right) \tag{58}$$

4. Inverse

If **A** and **D** are nonsingular $n \times n$ and $m \times m$ matrices, respectively, then

$$\begin{aligned}(\mathbf{A} + \mathbf{B}\mathbf{D}\mathbf{C})^{-1} \\ = \mathbf{A}^{-1} - \mathbf{A}^{-1}\mathbf{B}\left(\mathbf{C}\mathbf{A}^{-1}\mathbf{B} + \mathbf{D}^{-1}\right)^{-1}\mathbf{C}\mathbf{A}^{-1}\end{aligned} \tag{59}$$

Direct application of eq.(59) to control theory problems yields the following result,

$$\begin{aligned}\left[\mathbf{I} + \mathbf{C}(s\mathbf{I} - \mathbf{A})^{-1}\mathbf{B}\right]^{-1} \\ = \mathbf{I} - \mathbf{C}(s\mathbf{I} - \mathbf{A} + \mathbf{B}\mathbf{C})^{-1}\mathbf{B}\end{aligned} \tag{60}$$

5. If inverses for **A** and **D** exist, then

$$\begin{bmatrix} \mathbf{A} & 0 \\ \mathbf{C} & \mathbf{D} \end{bmatrix}^{-1} = \begin{bmatrix} \mathbf{A}^{-1} & 0 \\ -\mathbf{D}^{-1}\mathbf{C}\mathbf{A}^{-1} & \mathbf{D}^{-1} \end{bmatrix} \tag{61}$$

$$\begin{bmatrix} \mathbf{A} & \mathbf{B} \\ 0 & \mathbf{D} \end{bmatrix}^{-1} = \begin{bmatrix} \mathbf{A}^{-1} & -\mathbf{A}^{-1}\mathbf{B}\mathbf{D}^{-1} \\ 0 & \mathbf{D}^{-1} \end{bmatrix} \tag{62}$$

6. If the indicated inverses exist then,

$$\begin{bmatrix} \mathbf{A} & \mathbf{B} \\ 0 & \mathbf{D} \end{bmatrix}^{-1} = \begin{bmatrix} \mathbf{A}^{-1} & -\mathbf{A}^{-1}\mathbf{B}\mathbf{D}^{-1} \\ 0 & \mathbf{D}^{-1} \end{bmatrix} \tag{63}$$

where $\Delta_{\mathbf{A}}$ and $\Delta_{\mathbf{D}}$ are the Schur components of the matrices **A** and **D** defined by,

$$\begin{aligned}\Delta_{\mathbf{A}} &= \mathbf{A} - \mathbf{B}\mathbf{D}^{-1}\mathbf{C} \\ \Delta_{\mathbf{D}} &= \mathbf{D} - \mathbf{C}\mathbf{A}^{-1}\mathbf{B}\end{aligned} \tag{64}$$

Using property 1 the inverse of the Schur component $\Delta_{\mathbf{A}}$ in terms of $\Delta_{\mathbf{B}}$ is given by,

$$\Delta_A^{-1} = \mathbf{A}^{-1} + \mathbf{A}^{-1}\mathbf{B}\Delta_D^{-1}\mathbf{C}\mathbf{A}^{-1} \quad (65)$$

Resolvent Matrix

We have already seen that the characteristic polynomial a(λ) of any matrix **A** is det (λ**I** – **A**). The resolvent matrix is crucial in finding the impulse response of a system and is defined by $(\lambda\mathbf{I} - \mathbf{A})^{-1}$. This has bearing on the transfer function matrix in the state space formulation of systems.

32. SINGULAR VALUE DECOMPOSITION (SVD)

SVD Algorithm

One of the most elegant and efficient algorithms for the inversion of a matrix is the singular value decomposition (SVD). If an m × m square matrix **X** is of full rank, SVD is an efficient method for finding the inverse $\mathbf{X}^{-1}$. Even if **X** is an m × n rectangular matrix and is also not of full rank, then SVD is an elegant method of obtaining the rank and the pseudo inverse of **X**.

Any m × n rectangular matrix **X** whether full rank or otherwise can be decomposed into a diagonal matrix by two orthogonal matrices **U** and **V** as shown below:

$$\mathbf{U}^T \mathbf{X} \mathbf{V} = \begin{bmatrix} \Sigma_r & 0 \\ 0 & 0 \end{bmatrix} \tag{1}$$

where Σ_r is a diagonal matrix of r singular values of the matrix **X** given by,

$$\Sigma_r = \begin{bmatrix} \sigma_1 & 0 & \cdots & 0 \\ 0 & \sigma_2 & \cdots & 0 \\ \vdots & \vdots & \vdots & \vdots \\ 0 & 0 & \cdots & \sigma_r \end{bmatrix} \tag{2}$$

The rank of the matrix **X** is the number of nonzero singular values and in this case the rank is $r \leq \min(m, n)$.

The singular values are different from the eigenvalues of a square matrix discussed in the last section. The similarity transformation using the modal matrix **M** of eigenvectors of a square matrix **X** diagonalizes **X** to its eigenvalues as shown below:

$$\mathbf{M}^{-1} \mathbf{X} \mathbf{M} = \begin{bmatrix} \Lambda_r & 0 \\ 0 & 0 \end{bmatrix} \tag{3}$$

where Λ_r is the r-dimensional diagonal matrix of eigenvalues given by,

$$\Lambda_r = \begin{bmatrix} \lambda_1 & 0 & \cdots & 0 \\ 0 & \lambda_2 & \cdots & 0 \\ \vdots & \vdots & \vdots & \vdots \\ 0 & 0 & \cdots & \lambda_r \end{bmatrix} \tag{4}$$

The rank of the square matrix **X** is r. The modal matrix **M** is not orthogonal in general unless **X** is symmetric.

We shall now show the derivation of eq.(1) for the rectangular m × n matrix **X** with m > n. This case is called *overdetermined* where the numbers of rows are more than the number of columns.

Singular values $\{\sigma_1, \sigma_2, \ldots, \sigma_r\}$ of any non square m × n matrix **X** with m > n are the positive square roots of the eigenvalues of the symmetric Gram matrix $\mathbf{G} = \mathbf{X}^T\mathbf{X}$ described in Section 30. However, if the matrix **X** is symmetric and nonnegative definite, then its singular values are its eigenvalues.

We define the matrix **V** as the modal matrix of the Gram matrix **G**. Since **G** is symmetric and nonnegative definite, the modal matrix **V** is orthogonal. Hence we can write,

$$\mathbf{V}^T\mathbf{X}^T\mathbf{X}\,\mathbf{V} = \mathbf{V}^T\mathbf{G}\,\mathbf{V} = \begin{bmatrix} \Lambda_r & 0 \\ 0 & 0 \end{bmatrix} = \begin{bmatrix} \Sigma_r^2 & 0 \\ 0 & 0 \end{bmatrix} \tag{5}$$

with $\mathbf{V}^T\mathbf{V} = \mathbf{I}_n$ where $\mathbf{I}_n$ is an n-dimensional identity matrix. The n × n matrix **V** in eq.(5) can be partitioned into two submatrices $\mathbf{V}_1$, (n × r) and $\mathbf{V}_2$, (n × n–r).

$$\mathbf{V} = \left[\mathbf{V}_1 \middle| \mathbf{V}_2\right] \tag{6}$$

Using eq.(6) into eq.(5) we have,

$$\begin{bmatrix} \mathbf{V}_1^T \\ \hline \mathbf{V}_2^T \end{bmatrix} \mathbf{X}^T \mathbf{X} \left[\mathbf{V}_1 | \mathbf{V}_2 \right] = \begin{bmatrix} \Sigma_r^2 & 0 \\ 0 & 0 \end{bmatrix} \tag{7}$$

From eq.(7) we can write the following equations:

$$\begin{aligned} \mathbf{V}_1^T \mathbf{X}^T \mathbf{X} \mathbf{V}_1 &= \Sigma_r^2 \\ \mathbf{V}_2^T \mathbf{X}^T \mathbf{X} \mathbf{V}_2 &= 0 \end{aligned} \tag{8}$$

From eq.(8) we have,

$$\begin{aligned} \Sigma_r^{-1} \mathbf{V}_1^T \mathbf{X}^T \mathbf{X} \mathbf{V}_1 \Sigma_r^{-1} &= \mathbf{I} \\ \mathbf{X} \mathbf{V}_2 &= 0 \end{aligned} \tag{9}$$

If we now define in the first equation in eq.(9) an $m \times r$ matrix $\mathbf{U_1}$ such that,

$$\mathbf{U}_1 = \mathbf{X} \mathbf{V}_1 \Sigma_r^{-1} \tag{10}$$

then it is evident that

$$\mathbf{U}_1^T \mathbf{U}_1 = \mathbf{I}_r \tag{11}$$

We can define another $m \times m\text{–}r$ matrix $\mathbf{U_2}$ such that

$$\mathbf{U}_2^T \mathbf{U}_2 = \mathbf{I}_{m-r} , \mathbf{U}_1^T \mathbf{U}_2 = 0 , \mathbf{U}_2^T \mathbf{U}_1 = 0 \tag{12}$$

We form an $m \times m$ matrix $\mathbf{U}$ with $\mathbf{U_1}$ and $\mathbf{U_2}$, by setting $\mathbf{U} = [\mathbf{U}_1 | \mathbf{U}_2]$. $\mathbf{U}$ is an orthogonal matrix with $\mathbf{U}^T \mathbf{U} = \mathbf{I}_m$.

We have now found two orthogonal matrices $\mathbf{V}$ and $\mathbf{U}$ so that we can now write,

$$\begin{aligned} \mathbf{U}^T \mathbf{X} \mathbf{V} &= \begin{bmatrix} \mathbf{U}_1^T \\ \hline \mathbf{U}_2^T \end{bmatrix} \mathbf{X} \left[\mathbf{V}_1 | \mathbf{V}_2 \right] \\ &= \begin{bmatrix} \mathbf{U}_1^T \mathbf{X} \mathbf{V}_1 & \mathbf{U}_1^T \mathbf{X} \mathbf{V}_2 \\ \mathbf{U}_2^T \mathbf{X} \mathbf{V}_1 & \mathbf{U}_2^T \mathbf{X} \mathbf{V}_2 \end{bmatrix} \\ &= \begin{bmatrix} \Sigma_r^{-1} \mathbf{V}_1^T \mathbf{X}^T \mathbf{X} \mathbf{V}_1 & \mathbf{U}_1^T .0 \\ \mathbf{U}_2^T \mathbf{U}_1 \Sigma_r & \mathbf{U}_2^T .0 \end{bmatrix} \\ &= \begin{bmatrix} \Sigma_r^{-1} \Sigma_r^2 & 0 \\ 0 & 0 \end{bmatrix} \\ &= \begin{bmatrix} \Sigma_r & 0 \\ 0 & 0 \end{bmatrix} \end{aligned} \tag{13}$$

Eq.(13) is exactly eq.(1) and the decomposition is called the *singular value decomposition.*

A similar analysis can be performed for the *underdetermined* case where m < n. In this case, an analogous procedure as outlined previously with the outer product $\mathbf{XX}^T$ will give a similar singular value decomposition as given in eq.(13).

Conditioning Number

The ratio between the maximum and the minimum singular values of the full rank matrix **X** is known as the conditioning number $\chi(\mathbf{X})$ of **X**.

$$\chi(\mathbf{X}) = \frac{\sigma_{max}}{\sigma_{min}} \tag{14}$$

If this number is very large, then the matrix **X** is called ill-conditioned and conventional inversion techniques, (Gauss-Jordan, Gauss-Seidel, etc.) for square matrices will yield erroneous results. On the other hand, for SVD techniques the matrices need not even be square. For rank-deficient matrices the conditioning number χ is infinite.

The absolute value of the ratio of the maximum eigenvalue to the minimum eigenvalue will also give us an idea of the ill-conditioned nature of the matrix **X**. However, if **X** is symmetric and non-negative definite, this ratio is the same as the conditioning number χ.

Pseudo Inverse

We are now in a position to define a generalized inverse of a rectangular matrix called the pseudo inverse of **X**. It has meaning only if it becomes the actual inverse when the matrix **X** is square and full rank. The pseudo inverse of any m × n matrix **X** is defined by,

$$\mathbf{X}^{\dagger} = \begin{bmatrix} \left(\mathbf{X}^{\mathrm{T}}\,\mathbf{X}\right)^{-1}\mathbf{X}^{\mathrm{T}} & : & \mathrm{m} > \mathrm{n}\ \text{overdetermined} \\ \mathbf{X}^{\mathrm{T}}\left(\mathbf{X}\,\mathbf{X}^{\mathrm{T}}\right)^{-1} & : & \mathrm{m} < \mathrm{n}\ \text{underdetermined} \end{bmatrix} \tag{15}$$

$\mathbf{X}^{\dagger}$ has the same structure whether **X** is overdetermined or underdetermined. If m = n, then either one of eq.(15) can be used. Indeed, from the structure of the pseudo inverse when **X** is square and full rank $\mathbf{X}^{\dagger} = \mathbf{X}^{-1}$.

We explore the significance of the pseudo inverse. If the matrix **X** is square and full rank, then the set of linear equations given by $\mathbf{Ax} = \mathbf{y}$ has a unique solution $\mathbf{x} = \mathbf{A}^{-1}\mathbf{y}$. The inverse matrix $\mathbf{A}^{-1}$ exists because **A** is square and full rank. However, if **A** is not full rank, then we will not have a unique solution. An infinite number of solutions are possible if the system is overdetermined, that is, m > n. In this case the pseudo inverse gives that solution which has the minimum least square error.

We will now show how the pseudo inverse can be derived from the singular value decomposition. From the first equation in eq.(8) we can write,

$$\left(\mathbf{X}^{\mathrm{T}}\,\mathbf{X}\right)^{-1} = \mathbf{V}_1\,\Sigma_r^{-2}\,\mathbf{V}_1^{\mathrm{T}} \tag{16}$$

and from eq.(10) we can write,

$$\mathbf{X}^{\mathrm{T}} = \mathbf{V}_1\,\Sigma_r\,\mathbf{U}_1^{\mathrm{T}} \tag{17}$$

Substituting eqs.(16, 17) into the pseudo inverse eq.(15) for the overdetermined case we have,

$$\begin{aligned} \mathbf{X}^{\dagger} &= \left(\mathbf{X}^{\mathrm{T}}\,\mathbf{X}\right)^{-1}\mathbf{X}^{\mathrm{T}} \\ &= \left(\mathbf{V}_1\,\Sigma_r^{-2}\,\mathbf{V}_1^{\mathrm{T}}\right)\left(\mathbf{V}_1\,\Sigma_r\,\mathbf{U}_1^{\mathrm{T}}\right) \\ &= \mathbf{V}_1\,\Sigma_r^{-1}\,\mathbf{U}_1^{\mathrm{T}} \end{aligned}$$

$$= [\mathbf{V}_1\ \mathbf{V}_2]\begin{bmatrix}\Sigma_r^{-1} & 0\\ 0 & 0\end{bmatrix}\begin{bmatrix}\mathbf{U}_1^{\mathbf{T}}\\ \mathbf{U}_2^{\mathbf{T}}\end{bmatrix}$$
$$= \mathbf{V}\begin{bmatrix}\Sigma_r^{-1} & 0\\ 0 & 0\end{bmatrix}\mathbf{U}^{\mathbf{T}} \qquad (18)$$

If $\mathbf{X}$ is an m × m full rank matrix then $\mathbf{X}^\dagger$ is the regular inverse $\mathbf{X}^{-1}$ and is given by,

$$\mathbf{X}^{-1} = \mathbf{V}\,\Sigma_m^{-1}\,\mathbf{U}^{\mathbf{T}} \qquad (19)$$

where Σ_m is the diagonal matrix of all singular values of $\mathbf{X}$. Since $\mathbf{X}$ is full rank none of the singular values are 0. Eq.(19) involves the inverse of a diagonal matrix Σ_m that can be easily found by taking the reciprocal of its elements. $\mathbf{V}$ and $\mathbf{U}$ are orthogonal matrices that are found without inversion. They decompose $\mathbf{X}$ to its singular values as shown below:

$$\mathbf{U}^{\mathbf{T}}\,\mathbf{X}\,\mathbf{V} = \Sigma_m \qquad (20)$$

The method of finding the singular vectors $\mathbf{U}$ and $\mathbf{V}$ as shown here is for clarity of understanding. In actual practice, finding $\mathbf{U}$ and $\mathbf{V}$ from the matrix $\mathbf{X}^{\mathbf{T}}\mathbf{X}$ or $\mathbf{X}\mathbf{X}^{\mathbf{T}}$ is not very efficient. There are efficient algorithms like Jacobi rotations and Householder transformations for finding $\mathbf{U}$ and $\mathbf{V}$ directly from the matrix $\mathbf{X}$.

Example (Full Rank Square Matrix)

We shall first take the example given in eq.(16), Section 30, and apply SVD techniques to obtain the inverse.

$$\mathbf{X} = \begin{bmatrix} 1 & 3 & 5 & 4 & 1\\ 2 & -1 & -2 & -3 & 4\\ -1 & 4 & -4 & 2 & -5\\ 3 & 2 & 1 & 0 & 3\\ 4 & 0 & -3 & -1 & 2\end{bmatrix} \qquad (21)$$

The Gram matrix $\mathbf{G} = \mathbf{X}^T\mathbf{X}$ is shown below:

$$\mathbf{X}^T\mathbf{X} = \begin{bmatrix} 31 & 3 & -4 & -8 & 31 \\ 3 & 30 & 3 & 23 & -15 \\ -4 & 3 & 55 & 21 & 14 \\ -8 & 23 & 21 & 30 & -20 \\ 31 & -15 & 14 & -20 & 55 \end{bmatrix} \tag{22}$$

The eigenvalue matrix Λ_5 of the Gram matrix $\mathbf{G} = \mathbf{X}^T\mathbf{X}$ is given by,

$$\Lambda_5 = \begin{bmatrix} 90.753 & 0 & 0 & 0 & 0 \\ 0 & 37.681 & 0 & 0 & 0 \\ 0 & 0 & 2.679 & 0 & 0 \\ 0 & 0 & 0 & 0.261 & 0 \\ 0 & 0 & 0 & 0 & 69.626 \end{bmatrix} \tag{23}$$

The corresponding singular value matrix Σ_5 is given by the positive square root of eq.(23).

$$\Sigma_5 = \sqrt{\Lambda_5} = \begin{bmatrix} 9.526 & 0 & 0 & 0 & 0 \\ 0 & 6.138 & 0 & 0 & 0 \\ 0 & 0 & 1.637 & 0 & 0 \\ 0 & 0 & 0 & 0.511 & 0 \\ 0 & 0 & 0 & 0 & 8.344 \end{bmatrix} \tag{24}$$

Since none of the singular values are zero the matrix $\mathbf{X}$ is of full rank, namely 5. The conditioning number $\chi(\mathbf{X})$ is given by $9.526/0.511 = 18.64$, not an unreasonably high number.

The orthogonal right singular matrix $\mathbf{V}$ is the modal matrix of eigenvectors of $\mathbf{G}$. Or,

$$\mathbf{V} = \left[\phi_{G1}\ \phi_{G2}\ \phi_{G3}\ \phi_{G4}\ \phi_{G5}\right]^T \tag{25}$$

and it is computed as:

$$\mathbf{V} = \begin{bmatrix} 0.421 & 0.596 & -0.566 & -0.372 & 0.091 \\ -0.325 & 0.676 & 0.613 & -0.172 & 0.177 \\ -0.044 & -0.345 & 0.001 & -0.396 & 0.850 \\ -0.433 & 0.247 & -0.429 & 0.650 & 0.381 \\ 0.726 & 0.082 & 0.347 & 0.502 & 0.305 \end{bmatrix} \quad (26)$$

In the case of a full rank matrix **X** the left singular matrix **U** is computed from the equation,

$$\mathbf{U} = \mathbf{X}.\mathbf{V}\,\Sigma_5^{-1} \quad (27)$$

Substituting in eq.(27) for **X, V** and Σ_5^{-1} from eqs.(21), (3) and (24) respectively, we obtain the left singular matrix **U** as,

$$\mathbf{U} = \begin{bmatrix} -0.187 & 0.321 & -0.055 & 0.463 & 0.803 \\ 0.573 & 0.130 & 0.566 & 0.544 & -0.194 \\ -0.634 & 0.582 & 0.259 & 0.108 & 0.425 \\ 0.289 & 0.496 & 0.349 & -0.683 & 0.287 \\ 0.389 & 0.544 & -0.698 & 0.104 & -0.234 \end{bmatrix} \quad (28)$$

Since **X** is of full rank the inverse $\mathbf{X}^{-1}$ is given by,

$$\mathbf{X}^{-1} = \mathbf{V}.\Sigma_5^{-1}.\mathbf{U}^{\mathrm{T}}$$

$$= \begin{bmatrix} -0.287 & -0.556 & -0.145 & 0.441 & 0.233 \\ -0.118 & 0.020 & 0.137 & 0.412 & -0.255 \\ -0.294 & -0.451 & -0.157 & 0.529 & -0.137 \\ 0.662 & 0.515 & 0.103 & -0.941 & 0.309 \\ 0.463 & 0.694 & 0.105 & -0.559 & -0.017 \end{bmatrix} \quad (29)$$

Example (Rank Deficient Matrix)

The next example we consider is a 5×5 square singular matrix of rank 3 given in eq.(22) of Section 30.

$$\mathbf{X} = \begin{bmatrix} 6 & 11 & 5 & 4 & 1 \\ 5 & 3 & -2 & -3 & 4 \\ -8 & -12 & -4 & 2 & -5 \\ 6 & 7 & 1 & 0 & 3 \\ 3 & 0 & -3 & -1 & 2 \end{bmatrix} \tag{30}$$

The Gram matrix $\mathbf{G} = \mathbf{X}^T\mathbf{X}$ is shown on the next page.

$$\mathbf{G} = \mathbf{X}^{\mathbf{T}}\,\mathbf{X} = \begin{bmatrix} 170 & 219 & 49 & -10 & 90 \\ 219 & 323 & 104 & 11 & 104 \\ 49 & 104 & 55 & 21 & 14 \\ -10 & 11 & 21 & 30 & -20 \\ 90 & 104 & 14 & -20 & 55 \end{bmatrix} \tag{31}$$

The eigenvalue matrix Λ_3 of $\mathbf{G}$ is given by:

$$\Lambda_3 = \begin{bmatrix} 544.77 & 0 & 0 & 0 & 0 \\ 0 & 77.403 & 0 & 0 & 0 \\ 0 & 0 & 10.827 & 0 & 0 \\ 0 & 0 & 0 & 0 & 0 \\ 0 & 0 & 0 & 0 & 0 \end{bmatrix} \tag{32}$$

Unlike in eq.(23) where no eigenvalue was 0 we have in eq.(32) two eigenvalues of $\mathbf{G}$ are zero indicating thereby, that the rank of the matrix $\mathbf{X}$ is 5–2 = 3. The positive square root Σ_3 of Λ_3 is given by:

$$\Sigma_3 = \sqrt{\Lambda_3} = \begin{bmatrix} 23.34 & 0 & 0 & 0 & 0 \\ 0 & 8.798 & 0 & 0 & 0 \\ 0 & 0 & 3.29 & 0 & 0 \\ 0 & 0 & 0 & 0 & 0 \\ 0 & 0 & 0 & 0 & 0 \end{bmatrix} \tag{33}$$

Again the orthogonal right singular matrix $\mathbf{V}$ is the modal matrix of $\mathbf{G}$, given by,

$$\mathbf{V} = \begin{bmatrix} 0.5407 & -0.3363 & -0.4275 & -0.6234 & -0.1520 \\ 0.7651 & 0.2266 & 0.1014 & 0.5348 & -0.2586 \\ 0.2244 & 0.5630 & 0.5289 & -0.5348 & 0.2586 \\ 0.0046 & 0.5626 & -0.7120 & 0.0886 & 0.4106 \\ 0.2681 & -0.4495 & 0.1423 & 0.1772 & 0.8213 \end{bmatrix} \quad (34)$$

$\mathbf{V}$ can be partitioned into $[\mathbf{V}_1 \mid \mathbf{V}_2]$ where $\mathbf{V}_1$ and $\mathbf{V}_2$ are given by,

$$\mathbf{V}_1 = \begin{bmatrix} 0.5407 & -0.3363 & -0.4275 \\ 0.7651 & 0.2266 & 0.1014 \\ 0.2244 & 0.5630 & 0.5289 \\ 0.0046 & 0.5626 & -0.7120 \\ 0.2681 & -0.4495 & 0.1423 \end{bmatrix}$$

$$\mathbf{V}_2 = \begin{bmatrix} -0.6234 & -0.1520 \\ 0.5348 & -0.2586 \\ -0.5348 & 0.2586 \\ 0.0886 & 0.4106 \\ 0.1772 & 0.8213 \end{bmatrix} \quad (35)$$

with $\mathbf{V}_1^T\mathbf{V}_1 = \mathbf{I}_3$, $\mathbf{V}_2^T\mathbf{V}_2 = \mathbf{I}_2$
and $\mathbf{V}_1^T\mathbf{V}_2 = \mathbf{0}$, $\mathbf{V}_2^T\mathbf{V}_1 = \mathbf{0}$

The left singular matrix $\mathbf{U}$ can be determined by partitioning $\mathbf{U}$ as $\mathbf{U} = [\ \mathbf{U}_1 \mid \mathbf{U}_2\ ]$, and using eq.(10) to compute $\mathbf{U}_1$. Thus $\mathbf{U}_1$ is computed as,

$$\mathbf{U}_1 = \begin{bmatrix} -0.5599 & -0.5787 & 0.4592 \\ -0.2403 & 0.6380 & 0.0565 \\ 0.6742 & -0.1241 & 0.6223 \\ -0.4125 & 0.1383 & 0.2734 \\ -0.0634 & 0.4728 & 0.5691 \end{bmatrix} \quad (36)$$

$\mathbf{U}_2$ can be computed by using eqs.(12) and is found to be

$$\mathbf{U}_2 = \begin{bmatrix} -0.1663 & -0.3363 \\ 0.1761 & -0.7078 \\ 0.3118 & -0.2135 \\ 0.7204 & 0.4658 \\ -0.5702 & 0.3513 \end{bmatrix} \tag{37}$$

The complete left singular matrix **U** is given by:

$$\mathbf{U} = \begin{bmatrix} -0.5599 & -0.5787 & 0.4592 & -0.1663 & -0.3363 \\ -0.2403 & 0.6380 & 0.0565 & 0.1761 & -0.7078 \\ 0.6742 & -0.1241 & 0.6223 & 0.3118 & -0.2135 \\ -0.4125 & 0.1383 & 0.2734 & 0.7204 & 0.4658 \\ -0.0634 & 0.4728 & 0.5691 & -0.5702 & 0.3513 \end{bmatrix} \tag{38}$$

The singular value decomposition can be obtained from eqs.(1, 13) as,

$$\mathbf{U}^{\mathbf{T}}\,\mathbf{X}\,\mathbf{V} = \begin{bmatrix} \mathbf{U}_1 \\ \hline \mathbf{U}_2 \end{bmatrix} \mathbf{X} \left[\mathbf{V}_1 | \mathbf{V}_2 \right] = \begin{bmatrix} \Sigma_3 & 0 \\ 0 & 0 \end{bmatrix} \tag{39}$$

Since the rank of the 5 × 5 matrix **X** is only 3 we can not define an inverse for this matrix The pseudo inverse is obtained from eq.(18) and is given by,

$$\mathbf{X}^{\dagger} = \begin{bmatrix} -0.0505 & -0.0373 & -0.0605 & -0.0504 & -0.0935 \\ -0.0191 & 0.0103 & 0.0381 & -0.0015 & 0.0276 \\ 0.0314 & 0.0476 & 0.0986 & 0.0488 & 0.1211 \\ -0.1365 & 0.0285 & -0.1425 & -0.0504 & -0.0929 \\ 0.0430 & -0.0329 & 0.0410 & 0 & -0.0003 \end{bmatrix} \tag{40}$$

Note that unlike the regular inverse, neither $\mathbf{XX}^{\dagger}$ nor $\mathbf{X}^{\dagger}\mathbf{X}$ yield identity matrices.

33. VECTOR AND MATRIX DIFFERENTIATION

The definitions for differentiation with respect to a real vector and a complex vector are slightly different. In the case of differentiation of a complex vector the question of whether the derivative is a total or partial also makes a difference. We shall first discuss differentiation with respect to real vectors.

Differentiation with Respect to Real Vectors

Let **x** be an n-vector variable defined by $x = \{x_1, x_2, \ldots, x_n\}^T$. The total and partial derivative operator with respect to X is defined by

$$\frac{d}{d\mathbf{x}} = \begin{bmatrix} \dfrac{d}{dx_1} \\ \dfrac{d}{dx_2} \\ \vdots \\ \dfrac{d}{dx_n} \end{bmatrix} \quad \text{and} \quad \frac{\partial}{\partial \mathbf{x}} = \begin{bmatrix} \dfrac{\partial}{\partial x_1} \\ \dfrac{\partial}{\partial x_2} \\ \vdots \\ \dfrac{\partial}{\partial x_n} \end{bmatrix} \tag{1}$$

Analogous to $(d/dx)\, x = 1$ we want the derivative with respect to the vector **x** to be equal to the identity matrix **I**, taking care that the dimensional compatibility conditions are satisfied. Thus, $(d/d\mathbf{x})\, \mathbf{x}$ can not be defined since the multiplication $(n \times 1) \times (n \times 1)$ is not compatible. Therefore, we define $(d/d\mathbf{x}^T)\, \mathbf{x} = \mathbf{I}_n$ as shown below.

$$\frac{d}{d\mathbf{x}} \mathbf{x}^T = \begin{bmatrix} \dfrac{d}{dx_1} & \dfrac{d}{dx_2} & \cdots & \dfrac{d}{dx_n} \end{bmatrix}^T \begin{bmatrix} x_1 & x_2 & \cdots & x_n \end{bmatrix} = \mathbf{I}_n \tag{2}$$

where I_n is an $n \times n$ identity matrix defined by:

$$\mathbf{I_n} = \begin{bmatrix} 1 & 0 & \cdots & 0 \\ 0 & 1 & \cdots & 0 \\ \vdots & \vdots & \vdots & \vdots \\ 0 & 0 & \cdots & 1 \end{bmatrix}$$

In a similar fashion we can define the following operations

$$\frac{d}{d\mathbf{x}}\left(\mathbf{x}^T\mathbf{x}\right) = \begin{bmatrix} \frac{d}{dx_1} \\ \frac{d}{dx_2} \\ \vdots \\ \frac{d}{dx_n} \end{bmatrix} \left[x_1^2 + x_2^2 + \ldots + x_n^2\right] = \begin{bmatrix} 2x_1 \\ 2x_2 \\ \vdots \\ 2x_n \end{bmatrix} = \mathbf{Ix}+\mathbf{Ix} = 2\mathbf{x} \qquad (3)$$

We can also write the following differentiation formulae,

$$\frac{d}{d\mathbf{x}}\left(\mathbf{x}^T\mathbf{A}\,\mathbf{x}\right) = \frac{d}{d\mathbf{x}}\left(\mathbf{x}^T\mathbf{A}^T\mathbf{x}\right) = \left(\mathbf{A} + \mathbf{A}^T\right)\mathbf{x}$$

$$= 2\mathbf{Ax} \text{ if } \mathbf{A} \text{ is symmetric} \qquad (4)$$

$$\frac{d}{d\mathbf{x}}\left(\mathbf{y}^T\mathbf{A}\,\mathbf{x}\right) = \frac{d}{d\mathbf{x}}\left(\mathbf{x}^T\mathbf{A}^T\mathbf{y}\right) = \mathbf{A}^T\mathbf{y}$$

Note that in eq.(4) $\mathbf{x}^T\mathbf{Ax}$ is known as the quadratic form and is a scalar, and $\mathbf{y}^T\mathbf{Ax}$ is known as the bilinear form and is also a scalar. A scalar is its own transpose, that is, $\mathbf{x}^T\mathbf{Ax} = \mathbf{x}^T\mathbf{A}^T\mathbf{x}$ and $\mathbf{y}^T\mathbf{Ax} = \mathbf{x}^T\mathbf{A}^T\mathbf{y}$.

Analogous to eq.(2) the following operations are defined:

$$\left(\frac{d}{d\mathbf{x}}\right)^T \mathbf{x} = \left[\frac{d}{dx_1}\ \frac{d}{dx_2}\ \cdots\ \frac{d}{dx_n}\right] \begin{bmatrix} x_1 \\ x_2 \\ \vdots \\ x_n \end{bmatrix} = n \qquad (5)$$

$$\left(\frac{d}{d\mathbf{x}}\right)^T(\mathbf{x}\mathbf{x}^T) = \left[\frac{d}{dx_1}\ \frac{d}{dx_2} \cdots \frac{d}{dx_n}\right]\begin{bmatrix} x_1^2 & x_1x_2 & \cdots & x_1x_n \\ x_2x_1 & x_2^2 & \cdots & x_2x_n \\ \vdots & \vdots & \vdots & \vdots \\ x_nx_1 & x_nx_2 & \cdots & x_n^2 \end{bmatrix}$$

$$= (n + 1)\mathbf{x}^T$$

$$\left(\frac{d}{d\mathbf{x}}\right)^T(\mathbf{x}\mathbf{y}^T) = \left[\frac{d}{dx_1}\ \frac{d}{dx_2} \cdots \frac{d}{dx_n}\right]\begin{bmatrix} x_1y_1 & x_1y_2 & \cdots & x_1y_n \\ x_2y_1 & x_2y_2 & \cdots & x_2y_n \\ \vdots & \vdots & \vdots & \vdots \\ x_ny_1 & x_ny_2 & \cdots & x_ny_n \end{bmatrix}$$

$$= n\mathbf{y}^T \tag{6}$$

Example

Differentiate f(**x**) with respect to the vector **x**.

$$f(\mathbf{x}) = [x_1\ x_2]\begin{bmatrix} 1 & 2 \\ 3 & 4 \end{bmatrix}\begin{bmatrix} x_1 \\ x_2 \end{bmatrix} = x_1^2 + 5x_1x_2 + 4x_2^2 \tag{7}$$

Differentiating f(**x**) in eq.(7) with respect to x_1 and x_2 we obtain

$$\begin{aligned} \frac{d}{dx_1} f(\mathbf{x}) &= 2x_1 + 5x_2 \\ \frac{d}{dx_2} f(\mathbf{x}) &= 5x_1 + 8x_2 \end{aligned} = \begin{bmatrix} 2 & 5 \\ 5 & 8 \end{bmatrix}\begin{bmatrix} x_1 \\ x_2 \end{bmatrix} \tag{8}$$

Using the formula

$$\frac{df(\mathbf{x})}{d\mathbf{x}} = \frac{d(\mathbf{x}^T\mathbf{A}\mathbf{x})}{d\mathbf{x}} = [\mathbf{A} + \mathbf{A}^T]\mathbf{x}$$

we also obtain the same result:

$$[\mathbf{A} + \mathbf{A}^T]\,\mathbf{x} = \left\{\begin{bmatrix} 1 & 2 \\ 3 & 4 \end{bmatrix} + \begin{bmatrix} 1 & 3 \\ 2 & 4 \end{bmatrix}\right\}\begin{bmatrix} x_1 \\ x_2 \end{bmatrix} = \begin{bmatrix} 2 & 5 \\ 5 & 8 \end{bmatrix}\begin{bmatrix} x_1 \\ x_2 \end{bmatrix} \tag{9}$$

In a similar fashion, differentiating f(**x**, **y**) with respect to the vector **x** in the following example,

$$f(\mathbf{x},\mathbf{y}) = [y_1\ y_2]\begin{bmatrix}1 & 2\\ 3 & 4\end{bmatrix}\begin{bmatrix}x_1\\ x_2\end{bmatrix} = [x_1\ x_2]\begin{bmatrix}1 & 3\\ 2 & 4\end{bmatrix}\begin{bmatrix}y_1\\ y_2\end{bmatrix} \qquad (10)$$

$$= x_1y_1 + 3x_1y_2 + 2x_2y_1 + 4x_2y_2$$

we obtain

$$\begin{matrix}\frac{d}{dx_1} f(\mathbf{x},\mathbf{y}) = y_1 + 3y_2 \\ \frac{d}{dx_2} f(\mathbf{x},\mathbf{y}) = 2y_1 + 4y_2\end{matrix} = \begin{bmatrix}1 & 3\\ 2 & 4\end{bmatrix}\begin{bmatrix}y_1\\ y_2\end{bmatrix} \qquad (11)$$

Using the formula

$$\frac{df(\mathbf{x},\mathbf{y})}{d\mathbf{x}} = \frac{d(\mathbf{x}^T\mathbf{A}\mathbf{y})}{d\mathbf{x}} = \mathbf{A}^T\mathbf{y}$$

we can write,

$$\mathbf{A}^T\mathbf{y} = \begin{bmatrix}1 & 3\\ 2 & 4\end{bmatrix}\begin{bmatrix}y_1\\ y_2\end{bmatrix} = \begin{bmatrix}y_1 + 3y_2\\ 2y_1 + 4y_2\end{bmatrix} \qquad (12)$$

Complex Differentiation

Total Derivative

Differentiation with respect to a complex scalar variable $z = x + jy$ is a little bit more involved. If we have a function of a complex variable

$$w = f(z) = u(x, y) + jv(x, y) \qquad (13)$$

then the total derivative of $f(z)$ with respect to z is defined by

$$\frac{d}{dz} w = \lim_{\Delta z \to 0} \frac{\Delta w}{\Delta z} = \lim_{\Delta z \to 0} \frac{f(z + \Delta z) - f(z)}{\Delta z} \qquad (14)$$

where $\Delta z = \Delta x + j\Delta y$. If the derivative $(d/dz)\, z$ is to exist, then the limit must exist independent of the way in which Δx and Δy approach zero. This condition imposes restrictions on the existence of the total derivative. With $\Delta w = \Delta u + j\Delta v$ and $\Delta z = \Delta x + j\Delta y$ we have from eq.(14),

$$\lim_{\Delta z \to 0} \frac{\Delta w}{\Delta z} = \lim_{\substack{\Delta x \to 0 \\ \Delta y \to 0}} \frac{\Delta u + j\Delta v}{\Delta x + j\Delta y} \tag{15}$$

If we first let $\Delta y \to 0$ and then let $\Delta x \to 0$ in eq.(15), we have

$$\frac{df(z)}{dz} = \frac{\partial u}{\partial x} + j\frac{\partial v}{\partial x}$$

Similarly if we first let $\Delta x \to 0$ and then let $\Delta y \to 0$, we obtain,

$$\frac{df(z)}{dz} = \frac{\partial v}{\partial y} - j\frac{\partial u}{\partial y}$$

If the derivative $(d/dz)f(z)$ is to be unique, then these two results must be the same. Hence,

$$\frac{\partial u}{\partial x} = \frac{\partial v}{\partial y}$$
$$\frac{\partial v}{\partial x} = -\frac{\partial u}{\partial y} \tag{16}$$

These conditions are known as the Cauchy-Riemann (C-R) conditions. These are necessary and sufficient conditions for the existence of a total derivative of f(z).

A function f(z) is said to be an analytic function in the neighborhood of $z = z_0$, if C-R conditions are satisfied in that neighborhood. If it is also analytic at $z = z_0$, then the point z_0 is a regular point. However, if it is not analytic at $z = z_0$, then z_0 is called an isolated singular point or singular point.

Example

Let us find the total derivative of $f(z) = z^2$. $z^2 = (x + jy)^2 = x^2 - y^2 + 2jxy$. We first check whether f(z) satisfies the Cauchy–Riemann conditions of eq.(10).

$$\begin{aligned}
&z^2 = (x + jy)^2 = x^2 - y^2 + 2jxy \\
&u(x, y) = x^2 - y^2 : \;\; v(x, y) = 2xy \\
&\frac{\partial u}{\partial x} = 2x = \frac{\partial v}{\partial y} = 2x \\
&\frac{\partial v}{\partial x} = 2y = -\frac{\partial u}{\partial y} = 2y
\end{aligned} \qquad (17)$$

The C-R conditions are satisfied and hence

$$\frac{d}{dz} z^2 = \frac{\partial u}{\partial x} + j\frac{\partial v}{\partial x} = 2(x + jy) = 2z \qquad (18)$$

Example

On the other hand let us find the derivative of $f(z) = z^{*2} = (x - jy)^2 = x^2 - y^2 - 2jxy$. Here,

$$\begin{aligned}
&z^{*2} = (x - jy)^2 = x^2 - y^2 - 2jxy \\
&u(x, y) = x^2 - y^2 : \;\; v(x, y) = -2xy
\end{aligned} \qquad (19)$$

and

$$\frac{\partial u}{\partial x} = 2x \neq \frac{\partial v}{\partial y} = -2x$$

$$\frac{\partial v}{\partial x} = -2y \neq -\frac{\partial u}{\partial y} = 2y$$

Since the C-R conditions are not satisfied at points other than at $z = 0$, z^{*2} does not have a total derivative $(d/dz)\, z^{*2}$ at points other than at $z = 0$.

Partial Derivative

Many functions occurring in engineering practice are not analytic and we are interested in optimization problems involving the partial derivatives with respect to a complex variable z. With z = x + jy we shall define the partial derivative operator $(\partial/\partial z)$ as

$$\frac{\partial}{\partial z}=\frac{\partial}{\partial x}+j\frac{\partial}{\partial y} \tag{20}$$

Based on the definition given in eq.(20), we have the following.

$$\begin{aligned}\frac{\partial}{\partial z}z&=\left(\frac{\partial}{\partial x}+j\frac{\partial}{\partial y}\right)(x+jy)=1-1=0\\ \frac{\partial}{\partial z}z^*&=\left(\frac{\partial}{\partial x}+j\frac{\partial}{\partial y}\right)(x-jy)=1+1=2\end{aligned} \tag{21}$$

Note that the total derivative $(d/dz)\,z=1$ and $(d/dz)\,z^*$ does not exist since z* is not analytic except at z = 0.

From the definition $(\partial/\partial z)$ proposed in eq.(20) we can write the following equations,

$$\begin{aligned}\frac{\partial}{\partial z}z^*z&=\left(\frac{\partial}{\partial x}+j\frac{\partial}{\partial y}\right)(x^2+y^2)\\ &=2x+j2y=2z\\ \frac{\partial}{\partial z}z^2&=\left(\frac{\partial}{\partial x}+j\frac{\partial}{\partial y}\right)(x^2+2jxy-y^2)\\ &=2x+j2y-2x-j2y=0\\ \frac{\partial}{\partial z}(z^*w)&=\left(\frac{\partial}{\partial x}+j\frac{\partial}{\partial y}\right)\big(xu+j(xv-yu)+yv\big)\\ &=2(u+jv)=2w\\ \frac{\partial}{\partial z}(zw)&=\left(\frac{\partial}{\partial x}+j\frac{\partial}{\partial y}\right)\big(xu+j(xv+yu)-yv\big)=0\end{aligned} \tag{22}$$

If we used the chain rule of differentiation in eq.(22) and apply eq.(21), we obtain the same result. This fact gives credibility to our definition of the partial derivative,

$$\frac{\partial}{\partial z} = \frac{\partial}{\partial x} + j\,\frac{\partial}{\partial y}$$

On the other hand if z and w were real variables, then $(\partial/\partial z)\, z^2 = 2z$ and $(\partial/\partial z)\, zw = w$. These differences are accentuated in defining vector differentiation with respect to a complex variable vector **z**.

Vector Partial Differentiation

Analogous to eq.(1) we can define partial derivatives with respect to a vector complex variable **z**,

$$\mathbf{z} = [x_1 + jy_1,\ x_2 + jy_2,\ \ldots,\ x_n + jy_n]^T$$

as,

$$\frac{\partial}{\partial \mathbf{z}} = \begin{bmatrix} \frac{\partial}{\partial x_1} + j\,\frac{\partial}{\partial y_1} \\ \frac{\partial}{\partial x_2} + j\,\frac{\partial}{\partial y_2} \\ \vdots \\ \frac{\partial}{\partial x_n} + j\,\frac{\partial}{\partial y_n} \end{bmatrix} \qquad \frac{\partial}{\partial \mathbf{z}^*} = \begin{bmatrix} \frac{\partial}{\partial x_1} - j\,\frac{\partial}{\partial y_1} \\ \frac{\partial}{\partial x_2} - j\,\frac{\partial}{\partial y_2} \\ \vdots \\ \frac{\partial}{\partial x_n} - j\,\frac{\partial}{\partial y_n} \end{bmatrix} \tag{23}$$

With this definition of $(\partial/\partial z)$ and $(\partial/\partial z^*)$ we can define the following operations, eqs.(24-27).

$$\begin{aligned}\frac{\partial}{\partial \mathbf{z}}\mathbf{z}^{\mathbf{H}} &= \begin{bmatrix} \frac{\partial}{\partial x_1} + j\,\frac{\partial}{\partial y_1} \\ \vdots \\ \frac{\partial}{\partial x_n} + j\,\frac{\partial}{\partial y_n} \end{bmatrix} \left[x_1 - jy_1,\ \ldots\ ,\ x_n - jy_n\right] \\ &= \begin{bmatrix} 2 & 0 & 0 \\ \vdots & \vdots & \vdots \\ 0 & 0 & 2 \end{bmatrix} = 2\mathbf{I_n}\end{aligned} \tag{24}$$

where $\mathbf{z}^H$ is the conjugate transpose of $\mathbf{z}$ and $\mathbf{I_n}$ is an nth order identity matrix. In a similar manner,

$$\frac{\partial}{\partial \mathbf{z}^*}\mathbf{z}^T = \begin{bmatrix} \dfrac{\partial}{\partial x_1} - j\dfrac{\partial}{\partial y_1} \\ \vdots \\ \dfrac{\partial}{\partial x_n} - j\dfrac{\partial}{\partial y_n} \end{bmatrix} \left[x_1 + jy_1, \; \dots \; , x_n + jy_n\right]$$

$$= \begin{bmatrix} 2 & 0 & 0 \\ \vdots & \vdots & \vdots \\ 0 & 0 & 2 \end{bmatrix} = 2\mathbf{I_n} \tag{25}$$

In addition, we also have,

$$\frac{\partial}{\partial \mathbf{z}}\mathbf{z}^T = \begin{bmatrix} \dfrac{\partial}{\partial x_1} + j\dfrac{\partial}{\partial y_1} \\ \vdots \\ \dfrac{\partial}{\partial x_n} + j\dfrac{\partial}{\partial y_n} \end{bmatrix} \left[x_1 + jy_1, \; \dots \; , x_n + jy_n\right] \tag{26}$$

$$= \begin{bmatrix} 0 & 0 & 0 \\ \vdots & \vdots & \vdots \\ 0 & 0 & 0 \end{bmatrix} = 0_{\mathbf{n}}$$

and in an analogous manner

$$\frac{\partial}{\partial \mathbf{z}^*}\mathbf{z}^H = \begin{bmatrix} \dfrac{\partial}{\partial x_1} - j\dfrac{\partial}{\partial y_1} \\ \vdots \\ \dfrac{\partial}{\partial x_n} - j\dfrac{\partial}{\partial y_n} \end{bmatrix} \left[x_1 - jy_1, \; \dots \; , x_n - jy_n\right] \tag{27}$$

$$= \begin{bmatrix} 0 & 0 & 0 \\ \vdots & \vdots & \vdots \\ 0 & 0 & 0 \end{bmatrix} = 0_{\mathbf{n}}$$

Eqs.(24-27) are very useful in complex vector minimization of scalar quadratic error performance criteria as we will show in the following examples.

Example

z and **w** are two complex 2-vectors as shown below. It is desired to minimize the complex matrix scalar criteria $\varepsilon_1 = \mathbf{z}^H\mathbf{w}$ and $\varepsilon_2 = \mathbf{w}^H\mathbf{z}$ with respect to the complex vector **z**.

$$\mathbf{z} = \begin{bmatrix} a + j\,b \\ c + j\,d \end{bmatrix} \qquad \mathbf{w} = \begin{bmatrix} e + j\,f \\ g + j\,h \end{bmatrix} \tag{28}$$

and the corresponding error criteria are,

$$\begin{aligned} \varepsilon_1 &= \mathbf{z}^H\mathbf{w} = ae + bf + cg + dh + j(af - be + ch - dg) \\ \varepsilon_2 &= \mathbf{w}^H\mathbf{z} = ae + bf + cg + dh - j(af - be + ch - dg) \end{aligned} \tag{29}$$

Differentiating the error criteria ε_1 and ε_2 partially with respect to **z** we have,

$$\frac{\partial}{\partial \mathbf{z}} \varepsilon_1 = \frac{\partial}{\partial \mathbf{z}} \left(\mathbf{z}^H\mathbf{w}\right)$$

$$= \begin{bmatrix} \dfrac{\partial}{\partial a} + j\,\dfrac{\partial}{\partial b} \\ \dfrac{\partial}{\partial c} + j\,\dfrac{\partial}{\partial d} \end{bmatrix} \left[ae + bf + cg + dh + j\,(af - be + ch - dg)\right]$$

$$= \begin{bmatrix} e + e + j\,(f + f) \\ g + g + j\,(h + h) \end{bmatrix} = 2\mathbf{w}$$

$$\frac{\partial}{\partial \mathbf{z}} \varepsilon_2 = \frac{\partial}{\partial \mathbf{z}} \left(\mathbf{w}^H\mathbf{z}\right)$$

$$= \begin{bmatrix} \dfrac{\partial}{\partial a} + j\,\dfrac{\partial}{\partial b} \\ \dfrac{\partial}{\partial c} + j\,\dfrac{\partial}{\partial d} \end{bmatrix} \left[ae + bf + cg + dh - j\,(af - be + ch - dg)\right]$$

$$= \begin{bmatrix} e - e + j\,(f - f) \\ g - g + j\,(h - h) \end{bmatrix} = 0 \tag{30}$$

We can obtain the same result by the chain rule of differentiation and using eqs.(24-27) as shown below.

$$\begin{aligned} \frac{\partial}{\partial \mathbf{z}}\left(\mathbf{z}^H \mathbf{w}\right) &= \frac{\partial}{\partial \mathbf{z}} \mathbf{z}^H \mathbf{w} + \frac{\partial}{\partial \mathbf{z}} \mathbf{w}^T \mathbf{z}^* = 2\mathbf{w} + 0 = 2\mathbf{w} \\ \frac{\partial}{\partial \mathbf{z}}\left(\mathbf{w}^H \mathbf{z}\right) &= \frac{\partial}{\partial \mathbf{z}} \mathbf{w}^H \mathbf{z} + \frac{\partial}{\partial \mathbf{z}} \mathbf{z}^T \mathbf{w}^* = 0 + 0 = 0 \end{aligned} \tag{31}$$

Example

We take an example of minimizing a quadratic error surface with respect to a complex weight vector $\mathbf{a_p}$ and finding the optimum value of $\mathbf{a_p}$. The quadratic error surface is given by,

$$E_p = \sigma^2 + \mathbf{a_p^H R_p a_p} + \mathbf{r_p^H a_p} + \mathbf{a_p^H r_p} \tag{32}$$

We differentiate with respect to $\mathbf{a_p}$ and set the result to zero and solve for $\mathbf{a_{popt}}$ using eq.(26)

$$\begin{aligned} \frac{\partial}{\partial \mathbf{a_p}} E_p = \frac{\partial}{\partial \mathbf{a_p}}\left[\sigma^2 + \mathbf{a_p^H R_p a_p} + \mathbf{r_p^H a_p} + \mathbf{a_p^H r_p}\right] &= 0 \\ = \quad \frac{\partial}{\partial \mathbf{a_p}} \sigma^2 + \frac{\partial}{\partial \mathbf{a_p}} \mathbf{a_p^H R_p a_p} & \\ + \quad \frac{\partial}{\partial \mathbf{a_p}} \mathbf{a_p^T R_p a_p^*} + \frac{\partial}{\partial \mathbf{a_p}} \mathbf{a_p^T r_p^*} & \\ + \quad \frac{\partial}{\partial \mathbf{a_p}} \mathbf{a_p^H r_p} = 0 & \\ = \quad 0 + \mathbf{R_p a_p} + 0 + 0 + \mathbf{r_p} = 0 & \end{aligned} \tag{33}$$

From eq.(33) we can solve for $\mathbf{a_p}$ to yield $\mathbf{a_{popt}}$, given by

$$\mathbf{a_{popt}} = \mathbf{R_p^{-1} r_p}$$

The solution $\mathbf{a}_{\mathbf{popt}} = \mathbf{R}_{\mathbf{p}}^{-1}\mathbf{r}_{\mathbf{p}}$ is called the Wiener solution for the adaptive coefficients $\mathbf{a}_{\mathbf{p}}$.

In summary, we list the vector partial derivatives of real and complex quadratic forms.

For $\mathbf{x}$ real:

$$\frac{\partial}{\partial \mathbf{x}}\left(\mathbf{x}^{\mathbf{T}}\mathbf{A}\,\mathbf{x}\right) = \frac{\partial}{\partial \mathbf{x}}\,\mathbf{x}^{\mathbf{T}}\mathbf{A}\mathbf{x} + \frac{\partial}{\partial \mathbf{x}}\,\mathbf{x}^{\mathbf{T}}\mathbf{A}^{\mathbf{T}}\mathbf{x}$$

$$= \left(\mathbf{A} + \mathbf{A}^{\mathbf{T}}\right)\mathbf{x} \qquad (34)$$

$$\frac{\partial}{\partial \mathbf{x}}\left(\mathbf{x}^{\mathbf{T}}\mathbf{A}\,\mathbf{y}\right) = \mathbf{A}\,\mathbf{y}$$

For $\mathbf{x}$ complex:

$$\frac{\partial}{\partial \mathbf{x}}\left(\mathbf{x}^{\mathbf{H}}\mathbf{A}\,\mathbf{x}\right) = \frac{\partial}{\partial \mathbf{x}}\,\mathbf{x}^{\mathbf{H}}\mathbf{A}\mathbf{x} + \frac{\partial}{\partial \mathbf{x}}\,\mathbf{x}^{*}\mathbf{A}^{\mathbf{T}}\mathbf{x}^{\mathbf{T}}$$

$$= 2\mathbf{A}\mathbf{x} + 0 \qquad (35)$$

$$\frac{\partial}{\partial \mathbf{x}}\left(\mathbf{x}^{\mathbf{H}}\mathbf{A}\,\mathbf{y}\right) = 2\mathbf{A}\,\mathbf{y}$$

A final point to note in taking the vector partial derivatives is that if $\mathbf{x}$ is real $\left(\partial/\partial\mathbf{x}\right)\mathbf{x}^{\mathbf{T}} = \mathbf{I}_{\mathbf{n}}$ and if $\mathbf{x}$ is complex, $\left(\partial/\partial\mathbf{x}\right)\mathbf{x}^{\mathbf{T}} = 0$ and $\left(\partial/\partial\mathbf{x}\right)\mathbf{x}^{\mathbf{H}} = 2\mathbf{I}_{\mathbf{n}}$.

In addition, the matrix partial derivative with respect to a matrix $\mathbf{K}$ of the scalar $\mathrm{tr}\left(\mathbf{KAK}^{\mathrm{T}}\right)$ where $\mathbf{A}$ is a symmetric matrix is given by

$$\frac{\partial}{\partial \mathbf{K}}\,\mathrm{tr}\left(\mathbf{KAK}^{\mathrm{T}}\right) = 2\,\mathbf{AK} \qquad (36)$$

34. STATE SPACE TECHNIQUES

Concept of State (Continuous Time)

A linear system can be described by several characterizations. The familiar characterization is the input-output description as shown in diagram,

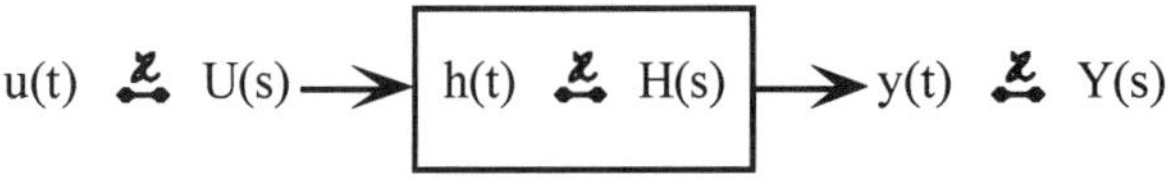

where, u(t) is the input time function, y(t) is the output time function and H(s) is the transfer function that determines the behavior of the system. Or,

$$Y(s) = H(s)U(s) \tag{1}$$

where U(s) is the Laplace Transform of the input, Y(s) is the Laplace Transform of the output and H(s) is independent of either the input or the output and expressed as a ratio of polynomials in s as shown below.

$$H(s) = \frac{b_m s^m + b_{m-1} s^{m-1} + \ldots + b_0}{s^n + a_{n-1} s^{n-1} + \ldots + a_0}, \; n > m \tag{2}$$

The linear system as shown in the figure above can also be described by an n-th order differential equation given by

$$\begin{aligned} &\frac{d^n y(t)}{dt^n} + a_{n-1}\frac{d^{n-1}y(t)}{dt^{n-1}} + \ldots + a_0 y(t) \\ &\quad = b_m \frac{d^m u(t)}{dt^m} + b_{m-1}\frac{d^{m-1}u(t)}{dt^{m-1}} + \ldots + b_0 u(t) \end{aligned} \tag{3}$$

If the input y(t) has to be bounded we require that the degree m be less than n in eqs.(2, 3). We also need n initial conditions $\{y(t_{0+}), y'(t_{0+}), \ldots, y^{(n-1)}(t_{0+})\}$ at $t = t_{0+}$, to solve for the trajectory y(t) uniquely. However, to obtain a complete description of the system we need to get not only the trajectory y(t) but also the trajectories for y'(t), y"(t), ... $y^{(n-1)}(t)$. This complete characterization of the dynamics of the system may be conveniently represented by the *state of the system* and a set of variables called *state variables*. For example, one set of state variables can be defined by $\{x_1(t) = y(t), x_2(t) = y'(t), x_3(t) = y''(t), \ldots x_n(t) = y^{(n-1)}(t)\}$. The general state variable characterization of a system is defined by the following two matrix equations:

$$\frac{d\mathbf{x}(t)}{dt} = \mathbf{A}\mathbf{x}(t) + \mathbf{B}\mathbf{u}(t) \qquad (4)$$
$$\mathbf{y}(t) = \mathbf{C}\mathbf{x}(t) + \mathbf{D}\mathbf{u}(t)$$

where **x(t)** is an $n \times 1$ vector of state variables, **u(t)** is an $m \times 1$ vector of control variables, **y(t)** is an $r \times 1$ vector of output variables, **A** is an $n \times n$ nonsingular matrix, **B** is an $n \times m$ matrix, **C** is an $r \times n$ matrix and **D** is an $r \times m$ matrix. The matrices **A, B, C, D** can be also be functions of time but we will restrict ourselves to the case where they are constants. We also have the initial condition matrix $(d\mathbf{x}/dt)_{0+}$. In defining the state variables, we should be careful to see that the derivatives are limited to the left-hand side of eqs.(3,4) and the right-hand side does not contain any derivatives. Note that state variable representations are not unique. We shall now discuss the three important configurations of state representation.

Formulation of State Equations

First Canonical Form (Controller Canonic Form)

Given the n-th order differential equation (3) under the assumption that initial conditions on y(t) and its derivatives are all zero, we can form the state variable representation. We can first write the following reduced differential equation containing only u(t) and none of its derivatives.

$$\frac{d^n w(t)}{dt^n} + a_{n-1}\frac{d^{n-1} w(t)}{dt^{n-1}} + \ldots + a_0 w(t) = u(t) \qquad (5)$$

We differentiate eq.(5) successively and form the sum,

$$b_m\frac{d^m u(t)}{dt^m} + b_{m-1}\frac{d^{m-1} u(t)}{dt^{m-1}} + \ldots + b_0 u(t)$$

resulting in the following equation,

$$\begin{aligned}
&\left[b_m\frac{d^m u(t)}{dt^m} + b_{m-1}\frac{d^{m-1} u(t)}{dt^{m-1}} + \ldots + b_0 u(t)\right] \\
&= \frac{d^n}{dt^n}\left[b_m\frac{d^m w(t)}{dt^m} + b_{m-1}\frac{d^{m-1} w(t)}{dt^{m-1}} + \ldots + b_0 w(t)\right] \\
&\quad + a_{n-1}\frac{d^{n-1}}{dt^{n-1}}\left[b_m\frac{d^m w(t)}{dt^m} + b_{m-1}\frac{d^{m-1} w(t)}{dt^{m-1}} + \ldots \right. \\
&\qquad\qquad \left. + b_0 w(t)\right] \qquad (6) \\
&\quad + \ldots \\
&\quad + a_0\left[b_m\frac{d^m w(t)}{dt^m} + b_{m-1}\frac{d^{m-1} w(t)}{dt^{m-1}} + \ldots + b_0 w(t)\right]
\end{aligned}$$

Comparing eq.(6) to eq.(3) we can write for the output variable y(t),

$$y(t) = \left[b_m\frac{d^m w(t)}{dt^m} + b_{m-1}\frac{d^{m-1} w(t)}{dt^{m-1}} + \ldots + b_0 w(t\right] \qquad (7)$$

For eq.(5), we can define the following state variables:

$$\begin{aligned}
x_1(t) &= w(t) \\
x_2(t) &= \frac{dx_1(t)}{dt} = \frac{dw(t)}{dt} \\
x_3(t) &= \frac{dx_2(t)}{dt} = \frac{d^2 w(t)}{dt^2} \qquad (8) \\
&\vdots \\
x_n(t) &= \frac{dx_{n-1}(t)}{dt} = \frac{d^{n-1} w(t)}{dt^{n-1}}
\end{aligned}$$

and the state equations for the differential equation (3) can be written as,

$$\frac{dx_n}{dt} = -a_{n-1}x_n(t) - a_{n-2}x_{n-1}(t) - \dots - a_0x_1(t) + u(t) \quad (9)$$

$$y(t) = b_m x_{m+1}(t) + b_{m-1}x_m(t) + \dots + b_0x_1(t) \quad (10)$$

The matrix formulation corresponding to eqs.(8, 10) can now be written as,

$$\frac{d}{dt}\left[x_1(t)\ \ x_2(t)\ \dots\ x_{n-1}(t)\ \ x_n(t)\right]^T$$

$$= \begin{bmatrix} 0 & 1 & \cdots & 0 \\ 0 & 0 & \cdots & 0 \\ \vdots & \vdots & \vdots & \vdots \\ 0 & 0 & \cdots & 1 \\ -a_0 & -a_1 & \cdots & -a_{n-1} \end{bmatrix} \begin{bmatrix} x_1(t) \\ x_2(t) \\ \vdots \\ x_{n-1}(t) \\ x_n(t) \end{bmatrix} + \begin{bmatrix} 0 \\ 0 \\ \vdots \\ 0 \\ 1 \end{bmatrix} u(t) \quad (11)$$

$$y(t) = \left[b_0\ \ b_1 \cdots b_m\ \ 0 \cdots 0\right] \begin{bmatrix} x_1(t) \\ x_2(t) \\ \vdots \\ x_{m+1}(t) \\ x_{m+2}(t) \\ \vdots \\ x_n(t) \end{bmatrix} \quad (12)$$

Eqs.(11, 12) can be expressed in the state space form corresponding to eq.(4) as follows.

$$\frac{d\mathbf{x}(t)}{dt} = \mathbf{A}_1\mathbf{x}(t) + \mathbf{B}_1u(t)$$

$$\mathbf{y}(t) = \mathbf{C}_1\mathbf{x}(t) \quad (13)$$

The signal flow diagram corresponding to eqs.(11, 12) is represented in the diagram on the next page.

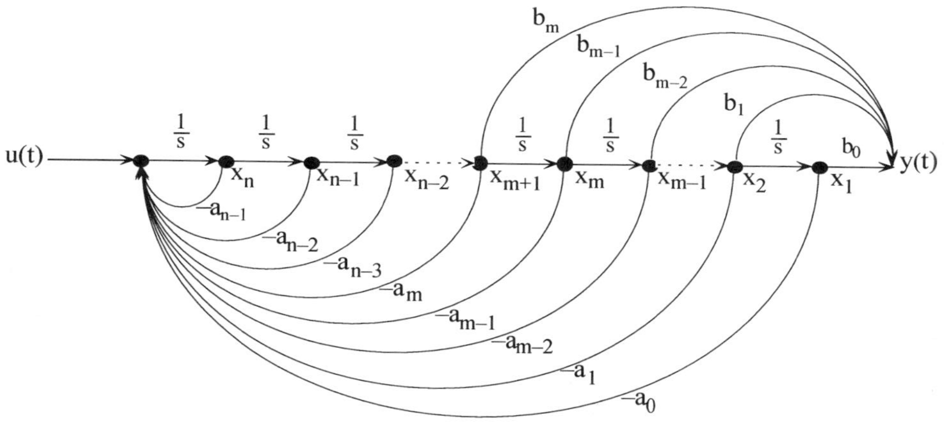

Signal Flow Diagram for Controller Canonic Form

Second Canonical Form (Observer Canonic Form)

In this canonical form we define the following state variables:

$$x_1(t) = y(t)$$

$$x_2(t) = \frac{d}{dt} y(t) + a_{n-1} y(t)$$

$$x_3(t) = \frac{d^2}{dt^2} y(t) + a_{n-1} \frac{d}{dt} y(t) + a_{n-2} y(t)$$

$$\vdots$$

$$x_{m+1}(t) = \frac{d^m}{dt^m} y(t) + a_{n-1} \frac{d^{m-1}}{dt^{m-1}} y(t) + \ldots + a_{n-m}\, y(t) - b_m\, u(t)$$

$$x_{m+2}(t) = \frac{d^{m+1}}{dt^{m+1}} y(t) + a_{n-1} \frac{d^m}{dt^m} y(t) + \ldots + a_{n-m-1} y(t) - b_m \frac{d}{dt} u(t) - b_{m-1}\, u(t) \tag{14}$$

$$\vdots$$

$$x_n(t) = \frac{d^{n-1}}{dt^{n-1}} y(t) + a_{n-1} \frac{d^{n-2}}{dt^{n-2}} y(t) + \ldots + a_1 y(t) - b_m \frac{d^{m-1}}{dt^{m-1}} u(t) - \ldots - b_1\, u(t)$$

If we differentiate the last equation in eq.(14) and substitute eq.(3) there results,

$$\frac{d}{dt} x_n(t) = \frac{d^n}{dt^n} y(t) + a_{n-1} \frac{d^{n-1}}{dt^{n-1}} y(t) + \ldots + a_1 \frac{d}{dt} y(t) - b_m \frac{d^m}{dt^m} u(t) - \ldots - b_1 \frac{d}{dt} u(t) \tag{15}$$
$$= -a_0\, y(t) + b_0 u(t) = -a_0\, x_1(t) + b_0 u(t)$$

Rearrangement of eq.(14) results in,

$$x_1(t) = y(t)$$

$$x_2(t) = \frac{d}{dt}x_1(t) + a_{n-1}y(t)$$

$$x_3(t) = \frac{d}{dt}x_2(t) + a_{n-2}y(t)$$

$$\vdots$$

$$x_{m+1}(t) = \frac{d}{dt}x_m(t) + \ldots + a_{n-m}y(t) - b_m u(t)$$

$$x_{m+2}(t) = \frac{d}{dt}x_{m+1}(t) + a_{n-m-1}y(t) - b_{m-1}\, u(t) \qquad (16)$$

$$\vdots$$

$$x_n(t) = \frac{d}{dt}x_{n-1}(t) + a_1 y(t) - b_1\, u(t)$$

Incorporating eq.(15) in eq.(16), and rearranging the terms in eq.(16) we obtain the state equations,

$$\frac{d}{dt}x_1(t) = -\,a_{n-1}x_1(t) + x_2(t)$$

$$\frac{d}{dt}x_2(t) = -\,a_{n-2}x_1(t) + x_3(t)$$

$$\vdots$$

$$\frac{d}{dt}x_m(t) = -\,a_{n-m}x_1(t) + x_{m+1}(t) + b_m u(t)$$

$$\frac{d}{dt}x_{m+1}(t) = -\,a_{n-m-1}x_1(t) + x_{m+2}(t) + b_{m-1}\, u(t) \qquad (17)$$

$$\vdots$$

$$\frac{d}{dt}x_{n-1}(t) = -\,a_1 x_1(t) + x_n(t) + b_1\, u(t)$$

$$\frac{d}{dt}x_n(t) = -a_0 x_1(t) + b_0 u(t)$$

$$y(t) = x_1(t) \qquad (18)$$

Equations (17, 18) can now be expressed as a matrix:

$$\frac{d}{dt}\left[x_1(t)\ \ x_2(t)\ \dots\ x_{m+1}(t)\ \ x_{m+2}(t)\ \dots\ x_{n-1}(t)\ \ x_n(t)\right]^T$$

$$= \begin{bmatrix} -a_{n-1} & 1\ 0 \dots 0\ 0 \dots 0 \\ -a_{n-2} & 0\ 1 \dots 0\ 0 \dots 0 \\ \vdots & \\ -a_{n-m} & 0\ 0 \dots 1\ 0 \dots 0 \\ -a_{n-m-1} & 0\ 0 \dots 0\ 1 \dots 0 \\ \vdots & \\ -a_1 & 0\ 0 \dots 0\ 0 \dots 1 \\ -a_0 & 0\ 0 \dots 0\ 0 \dots 0 \end{bmatrix} \begin{bmatrix} x_1(t) \\ x_2(t) \\ \vdots \\ x_{m+1}(t) \\ x_{m+2}(t) \\ \vdots \\ x_{n-1}(t) \\ x_n(t) \end{bmatrix} + \begin{bmatrix} 0 \\ 0 \\ \vdots \\ b_m \\ b_{m-2} \\ \vdots \\ b_1 \\ b_0 \end{bmatrix} \quad (19)$$

$$y(t) = \begin{bmatrix} 1 & 0 & 0 & \dots & 0 & 0 & \dots & 0 \end{bmatrix} \cdot \begin{bmatrix} x_1(t) \\ x_2(t) \\ \vdots \\ x_{m+1}(t) \\ x_{m+2}(t) \\ \vdots \\ x_{n-1}(t) \\ x_n(t) \end{bmatrix} \quad (20)$$

We can now write eqs.(19, 20) in the matrix notation corresponding to eq.(4):

$$\frac{d\mathbf{x}(t)}{dt} = \mathbf{A}_2 x(t) + \mathbf{B}_2 u(t)$$

$$\mathbf{y}(t) = \mathbf{C}_2 \mathbf{x}(t) \quad (21)$$

The signal flow graph representing the second canonical form is shown on the next page.

Third Canonical Form (Diagonal Canonic Form)

The previous two canonical forms are of the series type. This canonical form is of the parallel type. We assume that the characteristic equation given by the denominator of eq.(2) has distinct roots. In this case we

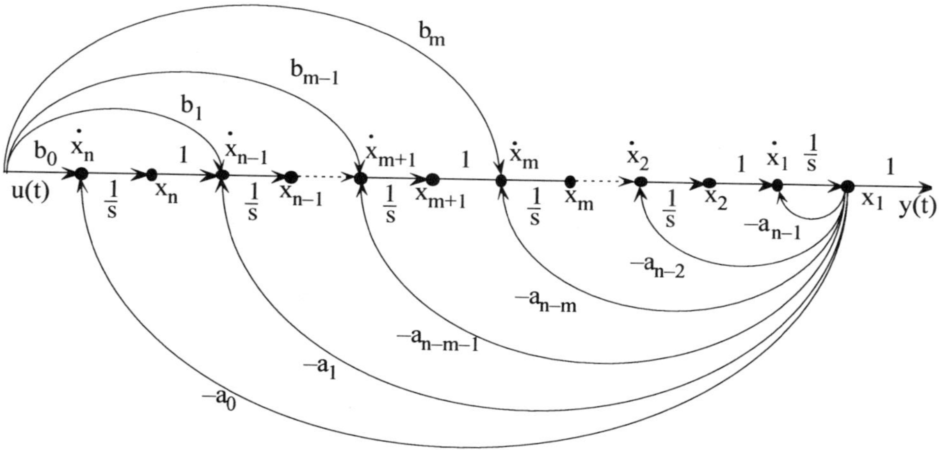

Signal Flow Diagram for Observer Canonic Form

can perform a partial fraction expansion for Y(s) as shown below.

$$Y(s) = \frac{\left[b_m s^m + b_{m-1} s^{m-1} + \ldots + b_0\right] U(s)}{s^n + a_{n-1} s^m + \ldots + a_0}$$
$$= \left[\frac{k_1}{s + s_1} + \frac{k_2}{s + s_2} + \ldots + \frac{k_n}{s + s_n}\right].U(s) \quad (22)$$

Each one of the terms on the right-hand side of eq.(22) represents a state variable as shown below.

$$X_i(s) = \frac{k_i U(s)}{s + s_i} \Leftrightarrow \frac{dx_i}{dt} + s_i x_i = k_i u(t),\ i = 1, 2, \ldots, n$$

Or the state variable representation of eq.(22) is,

$$\frac{dx_i}{dt} = -\,s_i x_i + k_i u(t),\ i = 1, 2, \ldots, n$$
$$y(t) = x_1(t) + x_2(t) + \ldots + x_n(t) \quad (23)$$

Eq.(23) can be represented by a signal flow diagram as follows:

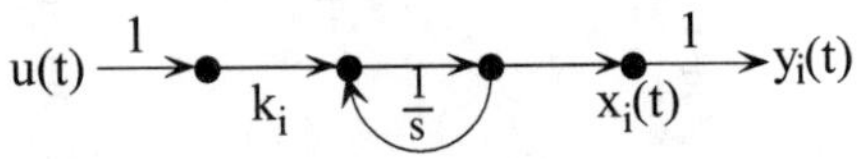

The complete state representation for the diagonal canonic form is,

$$\left[\frac{dx_1(t)}{dt}\ \frac{dx_2(t)}{dt} \cdots \frac{dx_i(t)}{dt} \cdots \frac{dx_n(t)}{dt}\right]^T$$
$$= \begin{bmatrix} -s_1 & 1 & \ldots & 0 & \ldots & 0 \\ 0 & -s_2 & \ldots & 0 & \ldots & 0 \\ \vdots & & & & & \\ 0 & 0 & \ldots & -s_i & \ldots & 0 \\ \vdots & & & & & \\ 0 & 0 & \ldots & 0 & \ldots & -s_n \end{bmatrix} \cdot \begin{bmatrix} x_1(t) \\ x_2(t) \\ \vdots \\ x_i(t) \\ \vdots \\ x_n(t) \end{bmatrix} + \begin{bmatrix} k_1 \\ k_2 \\ \vdots \\ k_i \\ \vdots \\ k_n \end{bmatrix} u(t) \quad (24)$$

$$y(t) = \begin{bmatrix} 1 & 1 & \dots & 1 & \dots & 0 \end{bmatrix} \cdot \begin{bmatrix} x_1(t) \\ x_2(t) \\ \vdots \\ x_i(t) \\ \vdots \\ x_n(t) \end{bmatrix} \tag{25}$$

We can now write eqs.(24, 25) in the matrix form corresponding to eq.(4)

$$\frac{d\mathbf{x}(t)}{dt} = \mathbf{A}_3 x(t) + \mathbf{B}_3 u(t)$$

$$\mathbf{y}(t) = \mathbf{C}_3 \mathbf{x}(t) \tag{26}$$

The signal flow diagram corresponding to eqs.(24, 25) is a parallel combination of the blocks similar to the one shown in the figure on the previous page and is drawn in the figure below.

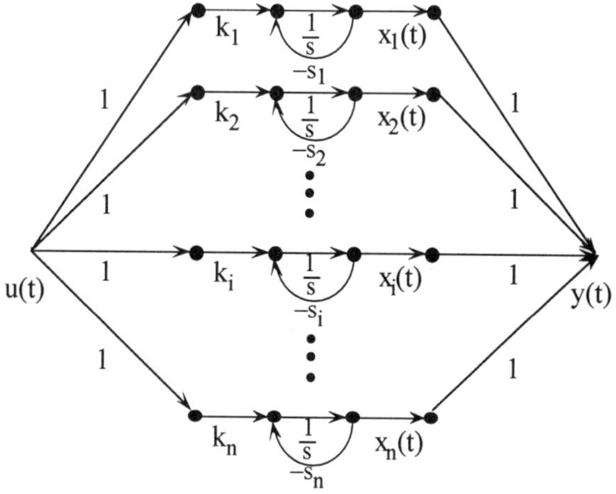

It is clear from the above discussions that the state variable representation is not unique and there can be any number of realizations. Since these realizations model the same differential equation (3), there must be a certain amount of commonality among these representations. In other words the matrices $\mathbf{A_i}$, $\mathbf{i}$ = 1, 2, 3, [eqs.(13, 21, 26)] and others like them are all similar. All these matrices have the same determinant $(-1)^n a_0$, and the same trace $-a_{n-1}$. The determinants of all the resolvent matrices $\Phi(s) = (s\mathbf{I} - \mathbf{A}_{i)})^{-1}$, i = 1, 2, 3 have the same characteristic polynomial, $s^n + a_{n-1}s^m + \ldots + a_0$. Therefore, we can apply similarity transformations to convert from one representation to another. The transformation matrix $\mathbf{T}$ is obtained by defining another state variable vector $\mathbf{z}$ by $\mathbf{x} = \mathbf{Tz}$ where $\mathbf{T}$ is nonsingular. Substituting $\mathbf{x} = \mathbf{Tz}$ and rearranging terms we can rewrite eq.(4) as,

$$\frac{d}{dt}\mathbf{z}(t) = \mathbf{T}^{-1}\mathbf{AT}\,\mathbf{z}(t) + \mathbf{T}^{-1}\mathbf{Bu}(t) \qquad (27)$$
$$\mathbf{y}(\mathbf{t}) = \mathbf{CT}\,\mathbf{z}(\mathbf{t}) + \mathbf{Du}(t)$$

In eqs.(13, 21), $\mathbf{A_1}$ and $\mathbf{A_2}$ are companion matrices that can be transformed into the diagonal form $\mathbf{A_3}$ by the transformations $\mathbf{T}_{13}{}^{-1}\mathbf{A}_1\mathbf{T}_{13} = \mathbf{A}_3$ and $\mathbf{T}_{23}{}^{-1}\mathbf{A}_2\mathbf{T}_{23} = \mathbf{A}_3$, where $\mathbf{T_{13}}$ and $\mathbf{T_{23}}$ are suitable Vandermonde matrices. These Vandermonde matrices $\mathbf{T_{13}}$ and $\mathbf{T}_{23}{}^{-1}$ corresponding to $\mathbf{A_1}$ and $\mathbf{A_2}$ are matrices of eigenvalues and their powers of $\mathbf{A_1}$ or $\mathbf{A_2}$ arranged in a suitable order as shown below.

$$\mathbf{T}_{13} = \begin{bmatrix} 1 & 1 & 1 & \cdots & 1 \\ \lambda_1 & \lambda_2 & \lambda_3 & \cdots & \lambda_n \\ \lambda_1^2 & \lambda_2^2 & \lambda_3^2 & \vdots & \lambda_n^2 \\ \vdots & \vdots & \vdots & \cdots & \vdots \\ \lambda_1^{n-1} & \lambda_2^{n-1} & \lambda_3^{n-1} & \cdots & \lambda_n^{n-1} \end{bmatrix}$$

$$
\mathbf{T}_{23}^{-1} = \begin{bmatrix} \lambda_1^{n-1} & \cdots & \lambda_1^2 & \lambda_1 & 1 \\ \lambda_2^{n-1} & \cdots & \lambda_2^2 & \lambda_2 & 1 \\ \lambda_2^{n-1} & \cdots & \lambda_2^2 & \lambda_2 & 1 \\ \vdots & \vdots & \vdots & \vdots & \vdots \\ \lambda_n^{n-1} & \cdots & \lambda_n^2 & \lambda_n & 1 \end{bmatrix} \tag{28}
$$

Similarly, to transform from $\mathbf{A_1}$ to $\mathbf{A_2}$ we rearrange the terms in the equation $\mathbf{T}_{13}^{-1}\mathbf{A}_1\mathbf{T}_{13} = \mathbf{T}_{23}^{-1}\mathbf{A}_2\mathbf{T}_{23}$ and write, $\mathbf{T}_{23}\mathbf{T}_{13}^{-1}\mathbf{A}_1\mathbf{T}_{13}\mathbf{T}_{23}^{-1} = \mathbf{A}_2$. As a result, the transformation matrix $\mathbf{T_{12}}$ is given by $\mathbf{T}_{12} = \mathbf{T}_{13}\mathbf{T}_{23}^{-1}$. Substituting eq.(28) in $\mathbf{T_{12}}$ and simplifying the result yields,

$$
\mathbf{T}_{12} = \begin{bmatrix} \sum_{k=1}^{n} \lambda_k^{n-1} & \cdots \sum_{k=1}^{n} \lambda_k^2 & \sum_{k=1}^{n} \lambda_k & \sum_{k=1}^{n} 1 \\ \sum_{k=1}^{n} \lambda_k^{n} & \cdots \sum_{k=1}^{n} \lambda_k^3 & \sum_{k=1}^{n} \lambda_k^2 & \sum_{k=1}^{n} \lambda_k \\ \sum_{k=1}^{n} \lambda_k^{n+1} & \cdots \sum_{k=1}^{n} \lambda_k^4 & \sum_{k=1}^{n} \lambda_k^3 & \sum_{k=1}^{n} \lambda_k^2 \\ \vdots & \vdots & \vdots & \vdots \\ \sum_{k=1}^{n} \lambda_k^{2(n-1)} & \cdots \sum_{k=1}^{n} \lambda_k^{n+1} & \sum_{k=1}^{n} \lambda_k^{n} & \sum_{k=1}^{n} \lambda_k^{n-1} \end{bmatrix} \tag{29}
$$

Example

Express the differential equation,

$$
\frac{d^4y(t)}{dt^4} + 10\frac{d^3y(t)}{dt^3} + 35\frac{d^2y(t)}{dt^2} + 40\frac{dy(t)}{dt} + 24\,y(t) = \frac{d^2u(t)}{dt^2} + 11\frac{du(t)}{dt} + 30u(t) \tag{30}
$$

with zero initial conditions in the three state variable canonic forms. The corresponding equation in the frequency domain is given by

$$Y(s) = \frac{(s+5)(s+6)\,U(s)}{(s+1)(s+2)(s+3)(s+4)} = \frac{\left(s^2+11s+30\right)U(s)}{s^4+10s^3+35s^2+40s+24} \qquad (31)$$

Since this is a fourth order differential equation we will have 4 state variables, $x_1(t)$, $x_2(t)$, $x_3(t)$, $x_4(t)$.

First Canonic Form

The state equations corresponding to eqs.(11, 12) are:

$$\frac{d}{dt}\begin{bmatrix} x_1(t) \\ x_2(t) \\ x_3(t) \\ x_4(t) \end{bmatrix} = \begin{bmatrix} 0 & 1 & 0 & 0 \\ 0 & 0 & 1 & 0 \\ 0 & 0 & 0 & 1 \\ -24 & -40 & -35 & -10 \end{bmatrix} \cdot \begin{bmatrix} x_1(t) \\ x_2(t) \\ x_3(t) \\ x_4(t) \end{bmatrix} + \begin{bmatrix} 0 \\ 0 \\ 0 \\ 1 \end{bmatrix} u(t)$$

$$y(t) = \begin{bmatrix} 30 & 11 & 1 & 0 \end{bmatrix} \cdot \begin{bmatrix} x_1(t) \\ x_2(t) \\ x_3(t) \\ x_4(t) \end{bmatrix} \qquad (32)$$

Second Canonic Form

The second canonic form can be obtained from eqs.(19, 20) as given below:

$$\frac{d}{dt}\begin{bmatrix} x_1(t) \\ x_2(t) \\ x_3(t) \\ x_4(t) \end{bmatrix} = \begin{bmatrix} -10 & 1 & 0 & 0 \\ -35 & 0 & 1 & 0 \\ -40 & 0 & 0 & 1 \\ -24 & 0 & 0 & 0 \end{bmatrix} \cdot \begin{bmatrix} x_1(t) \\ x_2(t) \\ x_3(t) \\ x_4(t) \end{bmatrix} + \begin{bmatrix} 0 \\ 1 \\ 11 \\ 30 \end{bmatrix} u(t)$$

$$y(t) = \begin{bmatrix} 1 & 0 & 0 & 0 \end{bmatrix} \cdot \begin{bmatrix} x_1(t) \\ x_2(t) \\ x_3(t) \\ x_4(t) \end{bmatrix} \qquad (33)$$

Third Canonic Form

This form is the parallel form that can be obtained from eqs.(24, 25). The partial fraction expansion of eq.(31) results in

$$Y(s) = \frac{(s+5)(s+6)}{(s+1)(s+2)(s+3)(s+4)}U(s)$$
$$= \left[-\frac{1}{3(s+4)} + \frac{3}{(s+3)} - \frac{6}{(s+2)} + \frac{10}{3(s+1)}\right]U(s) \tag{34}$$

Using eqs.(24, 25) the third canonic form can be written as,

$$\frac{d}{dt}\begin{bmatrix} x_1(t) \\ x_2(t) \\ x_3(t) \\ x_4(t) \end{bmatrix} = \begin{bmatrix} -4 & 0 & 0 & 0 \\ 0 & -3 & 0 & 0 \\ 0 & 0 & -2 & 0 \\ 0 & 0 & 0 & -1 \end{bmatrix} \cdot \begin{bmatrix} x_1(t) \\ x_2(t) \\ x_3(t) \\ x_4(t) \end{bmatrix} + \begin{bmatrix} -\frac{1}{3} \\ 3 \\ -6 \\ \frac{10}{3} \end{bmatrix} u(t)$$
$$y(t) = \begin{bmatrix} 1 & 1 & 1 & 1 \end{bmatrix} \cdot \begin{bmatrix} x_1(t) \\ x_2(t) \\ x_3(t) \\ x_4(t) \end{bmatrix} \tag{35}$$

From eqs.(32, 33, 35) the matrices $\mathbf{A}_1$, $\mathbf{A}_2$, $\mathbf{A}_3$ are defined below.

$$\mathbf{A}_1 = \begin{bmatrix} 0 & 1 & 0 & 0 \\ 0 & 0 & 1 & 0 \\ 0 & 0 & 0 & 1 \\ -24 & -40 & -35 & -10 \end{bmatrix} \quad \mathbf{A}_2 = \begin{bmatrix} -10 & 1 & 0 & 0 \\ -35 & 0 & 1 & 0 \\ -40 & 0 & 0 & 1 \\ -24 & 0 & 0 & 0 \end{bmatrix}$$

$$\mathbf{A}_3 = \begin{bmatrix} -4 & 0 & 0 & 0 \\ 0 & -3 & 0 & 0 \\ 0 & 0 & -2 & 0 \\ 0 & 0 & 0 & -1 \end{bmatrix} \tag{36}$$

The matrices $\mathbf{T_{13}}, \mathbf{T_{23}}, \mathbf{T_{12}}$ are given below:

$$\mathbf{T}_{13} = \begin{bmatrix} 1 & 1 & 1 & 1 \\ -4 & -3 & -2 & -1 \\ 16 & 9 & 4 & 1 \\ -64 & -27 & -8 & -1 \end{bmatrix} \quad \mathbf{T}_{23} = \begin{bmatrix} -\frac{1}{6} & \frac{1}{2} & -\frac{1}{2} & \frac{1}{6} \\ -1 & \frac{7}{2} & -4 & \frac{3}{2} \\ -\frac{11}{6} & 7 & -\frac{19}{2} & \frac{13}{3} \\ -1 & 4 & -6 & 4 \end{bmatrix}$$

$$\mathbf{T}_{12} = \mathbf{T}_{13}\mathbf{T}_{23}^{-1} = \begin{bmatrix} -100 & 30 & -10 & 4 \\ 354 & -100 & 30 & -10 \\ -1300 & 354 & -100 & 30 \\ 4890 & -1300 & 354 & -100 \end{bmatrix} \qquad (37)$$

The corresponding inverse matrices are,

$$\mathbf{T}_{13}^{-1} = \begin{bmatrix} -1 & -\frac{11}{6} & -1 & -\frac{1}{6} \\ 4 & 7 & \frac{7}{2} & \frac{1}{2} \\ -6 & -\frac{19}{2} & -4 & -\frac{1}{2} \\ 4 & \frac{13}{3} & \frac{3}{2} & \frac{1}{6} \end{bmatrix} \quad \mathbf{T}_{23}^{-1} = \begin{bmatrix} -64 & 16 & -4 & 1 \\ -27 & 9 & -3 & 1 \\ -8 & 4 & -2 & 1 \\ -1 & 1 & -1 & 1 \end{bmatrix}$$

$$\mathbf{T}_{12}^{-1} = \mathbf{T}_{23}\mathbf{T}_{13}^{-1} = \mathbf{T}_{21} = \begin{bmatrix} \frac{35}{6} & \frac{167}{18} & \frac{25}{6} & \frac{5}{9} \\ 45 & \frac{425}{6} & \frac{63}{2} & \frac{25}{6} \\ \frac{625}{6} & \frac{2905}{18} & \frac{425}{6} & \frac{167}{18} \\ 69 & \frac{625}{6} & 45 & \frac{35}{6} \end{bmatrix} \qquad (38)$$

We can now write $\begin{bmatrix} \mathbf{T}_{13}^{-1}\mathbf{A}_1\mathbf{T}_{13} = \mathbf{A}_3 & \mathbf{T}_{23}^{-1}\mathbf{A}_2\mathbf{T}_{23} = \mathbf{A}_3 \\ \mathbf{T}_{12}^{-1}\mathbf{A}_1\mathbf{T}_{12} = \mathbf{A}_2 & \mathbf{T}_{21}^{-1}\mathbf{A}_2\mathbf{T}_{21} = \mathbf{A}_1 \end{bmatrix}$

Finding State Transfer Matrix

We have formulated the state equations under zero initial conditions in the form,

$$\frac{d\mathbf{x}(t)}{dt} = \mathbf{A}\mathbf{x}(t) + \mathbf{B}\mathbf{u}(t)$$

$$\mathbf{y}(t) = \mathbf{C}\mathbf{x}(t) + \mathbf{D}\mathbf{u}(t) \qquad (4)$$

where $\mathbf{x}(t)$ is an n-vector, $\mathbf{u}(t)$ is an m-vector and $\mathbf{y}(t)$ is an r-vector. We will now take Laplace transforms on both sides of eq.(4) and write in the frequency domain,

$$s\mathbf{X}(s) = \mathbf{A}\mathbf{X}(s) + \mathbf{B}\mathbf{U}(s) \qquad (39a)$$

$$\mathbf{Y}(s) = \mathbf{C}\mathbf{X}(s) + \mathbf{D}\mathbf{U}(s) \qquad (39b)$$

The solution for $\mathbf{X}(s)$ is obtained by solving eq.(39a),

$$\mathbf{X}(s) = (s\mathbf{I} - \mathbf{A})^{-1}\mathbf{B}\mathbf{U}(s) = \Phi(s)\mathbf{B}\mathbf{U}(s) \qquad (40a)$$

Substituting eq.(40a) in eq.(39b) we have for $\mathbf{Y}(s)$,

$$\mathbf{Y}(s) = \left[\mathbf{C}(s\mathbf{I}-\mathbf{A})^{-1}\mathbf{B}+\mathbf{D}\right]\mathbf{U}(s) = \left[\mathbf{C}\Phi(s)\mathbf{B}+\mathbf{D}\right]\mathbf{U}(s) \quad (40b)$$

As noted earlier, the resolvent of the matrix $\mathbf{A}$ in eqs.(40) is $\Phi(s) = (s\mathbf{I} - \mathbf{A})^{-1}$.

Equation (40a) expresses the matrix transfer relationship of the state n-vector $\mathbf{X}(s)$ to the input m-vector $\mathbf{U}(s)$. Equation (40b) expresses the matrix transfer relationship of the output r-vector $\mathbf{Y}(s)$ to the input m-vector $\mathbf{U}(s)$. The matrix $\Phi(s) = (s\mathbf{I} - \mathbf{A})^{-1}$ can be expressed as,

$$\Phi(s) = \left[\ \phi_{.1}(s)\ \phi_{.2}(s)\ \cdots\ \phi_{.j}(s)\ \cdots\ \phi_{.n}(s)\ \right]$$

$$= \begin{bmatrix} \phi_{11}(s) & \phi_{12}(s) & \cdots & \phi_{1j}(s) & \cdots & \phi_{1n}(s) \\ \phi_{21}(s) & \phi_{22} & \cdots & \phi_{2j}(s) & \cdots & \phi_{2n}(s) \\ \vdots & \vdots & \vdots & \vdots & \vdots & \vdots \\ \phi_{i1}(s) & \phi_{i2} & \cdots & \phi_{ij}(s) & \cdots & \phi_{in}(s) \\ \vdots & \vdots & \vdots & \vdots & \vdots & \vdots \\ \phi_{n1}(s) & \phi_{n2} & \cdots & \phi_{nj}(s) & \cdots & \phi_{nn}(s) \end{bmatrix} \qquad (41)$$

Equation (40a) can be expanded using eq.(41) to yield,

$$\begin{bmatrix} X_1(s) \\ X_2(s) \\ \vdots \\ X_i(s) \\ \vdots \\ X_n(s) \end{bmatrix} = \begin{bmatrix} \phi_{11}(s) & \phi_{12}(s) & \cdots & \phi_{1j}(s) & \cdots & \phi_{1n}(s) \\ \phi_{21}(s) & \phi_{22} & \cdots & \phi_{2j}(s) & \cdots & \phi_{2n}(s) \\ \vdots & \vdots & \vdots & \vdots & \vdots & \vdots \\ \phi_{i1}(s) & \phi_{i2} & \cdots & \phi_{ij}(s) & \cdots & \phi_{in}(s) \\ \vdots & \vdots & \vdots & \vdots & \vdots & \vdots \\ \phi_{n1}(s) & \phi_{n2} & \cdots & \phi_{nj}(s) & \cdots & \phi_{nn}(s) \end{bmatrix}$$

$$\times . \begin{bmatrix} b_{11} & b_{12} & \cdots & b_{1j} & \cdots & b_{1m} \\ b_{21} & b_{22} & \cdots & b_{2j} & \cdots & b_{2m} \\ \vdots & \vdots & \vdots & \vdots & \vdots & \vdots \\ b_{i1} & b_{i2} & \cdots & b_{ij} & \cdots & b_{im} \\ \vdots & \vdots & \vdots & \vdots & \vdots & \vdots \\ b_{n1} & b_{n2} & \cdots & b_{nj} & \cdots & b_{nm} \end{bmatrix} . \begin{bmatrix} U_1(s) \\ U_2(s) \\ \vdots \\ U_j(s) \\ \vdots \\ U_m(s) \end{bmatrix}$$

$$= \begin{bmatrix} H_{11}(s) & H_{12}(s) & \ldots & H_{1j}(s) & \ldots & H_{1m}(s) \\ H_{21}(s) & H_{22}(s) & \ldots & H_{2j}(s) & \ldots & H_{2m}(s) \\ \vdots & & & & & \\ H_{i1}(s) & H_{i2}(s) & \ldots & H_{ij}(s) & \ldots & H_{im}(s) \\ \vdots & & & & & \\ H_{n1}(s) & H_{n2}(s) & \ldots & H_{nj}(s) & \ldots & H_{nm}(s) \end{bmatrix} . \begin{bmatrix} U_1(s) \\ U_2(s) \\ \vdots \\ U_j(s) \\ \vdots \\ U_m(s) \end{bmatrix} \quad (42)$$

Using eq.(42) in eq.(40b), Y(s) can be written as:

$$\begin{bmatrix} Y_1(s) \\ Y_2(s) \\ \vdots \\ Y_i(s) \\ \vdots \\ Y_r(s) \end{bmatrix} = \begin{bmatrix} c_{11}(s) & c_{12}(s) & \ldots & c_{1j}(s) & \ldots & c_{1m}(s) \\ c_{21}(s) & c_{22}(s) & \ldots & c_{2j}(s) & \ldots & c_{2m}(s) \\ \vdots & & & & & \\ c_{i1}(s) & c_{i2}(s) & \ldots & c_{ij}(s) & \ldots & c_{im}(s) \\ \vdots & & & & & \\ c_{n1}(s) & c_{n2}(s) & \ldots & c_{nj}(s) & \ldots & c_{nm}(s) \end{bmatrix} \times$$

$$
\times . \begin{bmatrix} H_{11}(s) & H_{12}(s) \dots & H_{1j}(s) \dots & H_{1m}(s) \\ H_{21}(s) & H_{22}(s) \dots & H_{2j}(s) \dots & H_{2m}(s) \\ \vdots & & & \\ H_{i1}(s) & H_{i2}(s) \dots & H_{ij}(s) \dots & H_{im}(s) \\ \vdots & & & \\ H_{n1}(s) & H_{n2}(s) \dots & H_{nj}(s) \dots & H_{nm}(s) \end{bmatrix} \cdot \begin{bmatrix} U_1(s) \\ U_2(s) \\ \vdots \\ U_j(s) \\ \vdots \\ U_m(s) \end{bmatrix}
$$

$$
+ \begin{bmatrix} d_{11}(s) & d_{12}(s) \dots & d_{1j}(s) \dots & d_{1m}(s) \\ d_{21}(s) & d_{22}(s) \dots & d_{2j}(s) \dots & d_{2m}(s) \\ \vdots & & & \\ d_{i1}(s) & d_{i2}(s) \dots & d_{ij}(s) \dots & d_{im}(s) \\ \vdots & & & \\ d_{n1}(s) & d_{n2}(s) \dots & d_{nj}(s) \dots & d_{nm}(s) \end{bmatrix} \begin{bmatrix} U_1(s) \\ U_2(s) \\ \vdots \\ U_j(s) \\ \vdots \\ U_m(s) \end{bmatrix} \quad (43)
$$

The inverse Laplace transform of $X_i(s)$ and $Y_i(s)$ gives the output $x_i(t)$ and $y_i(t)$ respectively. The inverse Laplace transforms can be obtained by either the partial fraction expansion or contour integration techniques.

Example

We shall now find the transfer matrix for the three canonic forms found earlier in eqs.(32, 33, 35). The determinants associated with the three resolvent matrices $(s\mathbf{I} - \mathbf{A}_1)^{-1}$, $(s\mathbf{I} - \mathbf{A}_2)^{-1}$, $(s\mathbf{I} - \mathbf{A}_3)^{-1}$ are the characteristic equation $s^4 + 10s^3 + 35s^2 + 40s + 24$.

First Canonic Form

$$
\mathbf{A}_1 = \begin{bmatrix} 0 & 1 & 0 & 0 \\ 0 & 0 & 1 & 0 \\ 0 & 0 & 0 & 1 \\ -24 & -40 & -35 & -10 \end{bmatrix}
$$

$$
[\,s\mathbf{I} - \mathbf{A}_1]^{-1} = \begin{bmatrix} s & -1 & 0 & 0 \\ 0 & s & -1 & 0 \\ 0 & 0 & s & -1 \\ 24 & 40 & 35 & s+10 \end{bmatrix}^{-1}
$$

$$
= \frac{1}{s^4 + 10s^3 + 35s^2 + 40s + 24} \times
$$

$$
\times \begin{bmatrix} s^3+10s^2+35s+50 & s^2+10s+35 & s+10 & 1 \\ -24 & s(s^2+10s+35) & s(s+10) & s \\ -24s & -2(25s+12) & s^2(s+10) & s^2 \\ -24s^2 & -2s(25s+12 & -(35s^2+50s+24) & s^3 \end{bmatrix} \quad (44)
$$

The state frequency function matrix **X**(s) can be computed by substituting eq.(44) into eq.(40a) with **B** being a column vector $[0\ 0\ 0\ 1]^T$ given in eq.(32). Or,

$$
X(s) = \frac{1}{s^4 + 10s^3 + 35s^2 + 40s + 24} \begin{bmatrix} 1 \\ s \\ s^2 \\ s^3 \end{bmatrix} U(s) \quad (45)
$$

and the scalar output Y(s) is given by,

$$
Y(s) = \frac{[30\ 11\ 1\ \ 0\,]}{s^4 + 10s^3 + 35s^2 + 40s + 24} \begin{bmatrix} 1 \\ s \\ s^2 \\ s^3 \end{bmatrix} U(s)
$$

$$
= \frac{\left(s^2+11s+30\right) U(s)}{s^4 + 10s^3 + 35s^2 + 40s + 24} \quad (46)
$$

Second Canonic Form

$$
A_2 = \begin{bmatrix} -10 & 1 & 0 & 0 \\ -35 & 0 & 1 & 0 \\ -40 & 0 & 0 & 1 \\ -24 & 0 & 0 & 0 \end{bmatrix}
$$

$$[\,sI - A_2]^{-1} = \begin{bmatrix} s+10 & -1 & 0 & 0 \\ 35 & s & -1 & 0 \\ 40 & 0 & s & -1 \\ 24 & 0 & 0 & s \end{bmatrix}^{-1}$$

$$= \frac{1}{s^4 + 10s^3 + 35s^2 + 40s + 24} \times$$

$$\times \begin{bmatrix} s^3 & s^2 & s & 1 \\ -(35s^2+50s+24) & s^2(s+10) & s(s+10) & s+10 \\ -2s(25s+12) & -2(25s+12) & s(s^2+10s+35) & s^2+10s+35 \\ -24s^2 & -24s & -24 & s^3+10s^2+35s+50 \end{bmatrix} \tag{47}$$

The state frequency function matrix **X**(s) can be computed by substituting eq.(47) into eq.(40a) with **B** being the column vector $[0\ 1\ 11\ 30]^T$ given in eq.(33). Or,

$$\mathbf{X}(s) = (\,s\mathbf{I} - \mathbf{A})^{-1}\mathbf{B}U(s)$$

$$= \frac{1}{s^4 + 10s^3 + 35s^2 + 40s + 24} \tag{48}$$

$$\times \begin{bmatrix} s^2 + 11s + 30 \\ s^3 + 21s^2 + 140s + 300 \\ 11s^3 + 140s^2 + 635s + 1026 \\ 6\left(5s^3 + 50s^2 + 171s + 206\right) \end{bmatrix} .U(s)$$

and the scalar output Y(s) is given by

$$Y(s) = \frac{[1 \ 0 \ 0 \ 0]}{s^4 + 10s^3 + 35s^2 + 40s + 24}$$

$$\times \begin{bmatrix} s^2 + 11s + 30 \\ s^3 + 21s^2 + 140s + 300 \\ 11s^3 + 140s^2 + 635s + 1026 \\ 6\left(5s^3 + 50s^2 + 171s + 206\right) \end{bmatrix} .U(s) \quad (49)$$

$$= \frac{\left(s^2+11s+30\right) U(s)}{s^4 + 10s^3 + 35s^2 + 40s + 24}$$

Third Canonic Form

$$\mathbf{A}_3 = \begin{bmatrix} -4 & 0 & 0 & 0 \\ 0 & -3 & 0 & 0 \\ 0 & 0 & -2 & 0 \\ 0 & 0 & 0 & -1 \end{bmatrix} \left[s\mathbf{I} - \mathbf{A}_3\right]^{-1} = \begin{bmatrix} \frac{1}{s+4} & 0 & 0 & 0 \\ 0 & \frac{1}{s+3} & 0 & 0 \\ 0 & 0 & \frac{1}{s+2} & 0 \\ 0 & 0 & 0 & \frac{1}{s+4} \end{bmatrix} \quad (50)$$

The state frequency function matrix $\mathbf{X}(s)$ can be computed by substituting eq.(50) into eq.(40a) with $\mathbf{B}$ being a column vector $[-1/3 \ 3 \ -6 \ 10/3]^T$ given by eq.(35). Or,

$$\begin{bmatrix} X_1(s) \\ X_2(s) \\ X_3(s) \\ X_4(s) \end{bmatrix} = \begin{bmatrix} \frac{1}{s+4} & 0 & 0 & 0 \\ 0 & \frac{1}{s+3} & 0 & 0 \\ 0 & 0 & \frac{1}{s+2} & 0 \\ 0 & 0 & 0 & \frac{1}{s+1} \end{bmatrix} \begin{bmatrix} -\frac{1}{3} \\ 3 \\ -6 \\ \frac{10}{3} \end{bmatrix} = \begin{bmatrix} -\frac{1}{3(s+4)} \\ \frac{3}{(s+3)} \\ -\frac{6}{(s+2)} \\ \frac{10}{(s+1)} \end{bmatrix} \quad (51)$$

The scalar output function Y(s) is given by:

$$Y(s) = [1\ 1\ 1\ 1] \begin{bmatrix} \dfrac{1}{3(s+4)} \\ \dfrac{3}{(s+3)} \\ \dfrac{6}{(s+2)} \\ \dfrac{10}{(s+1)} \end{bmatrix} = \frac{(s^2+11s+30)\,U(s)}{s^4 + 10s^3 + 35s^2 + 40s + 24} \tag{52}$$

Note that we have assumed in all the three cases that all initial conditions are zero.

Solving for the Time Function

Time Domain Method

The final step in the state formulation is solving for the trajectories of the state variable vector $\mathbf{x}(t)$ from the state frequency matrix $\mathbf{X}(s)$. There are several methods of determining the time function. We shall discuss some of the time domain and frequency domain techniques. The easier method is the frequency domain technique where we find the partial fraction expansion for each of the elements of the transfer function matrix $\Phi(s)$ and find the inverse Laplace transform by a table look up.

We shall first take the homogeneous matrix differential equation given by

$$\frac{d\mathbf{x}(t)}{dt} = \mathbf{A}\mathbf{x}(t) \tag{53}$$

and find the matrix impulse response with zero initial conditions. By analogy to the scalar case we can find the solution as $\Phi(t) = e^{\mathbf{A}t}$, $t > 0$. This is a matrix exponential known as the fundamental matrix or the *state transition matrix*. It can be given as an infinite series expansion as,

$$e^{\mathbf{A}t} = \mathbf{I} + \mathbf{A}t + \frac{\mathbf{A}^2t^2}{2!} + \frac{\mathbf{A}^3t^3}{3!} + \ldots + \frac{\mathbf{A}^{n-1}t^{n-1}}{(n-1)!} + \frac{\mathbf{A}^nt^n}{n!} \ldots + \ldots \tag{54}$$

Finding the state transition matrix from eq.(54) is not very useful since it will not give individual exponential terms but will give an overall series expansion. In order to obtain recognizable exponentials we need to express eq.(54) as a finite series. Such a representation is available using the Cayley-Hamilton theorem. We shall assume that the characteristic equation,

$$s^n + a_{n-1}s^{n-1} + \ldots + a_1 s + a_0 = 0,$$

of the matrix **A** has distinct eigenvalues. From the Cayley-Hamilton theorem that the matrix **A** satisfies its characteristic equation the following equation can be written.

$$\mathbf{A}^n = -\left(a_{n-1}\mathbf{A}^{n-1} + \ldots + a_1\mathbf{A} + a_0\mathbf{I}\right) \tag{55}$$

We can multiply both sides of the eq.(55) by $\mathbf{A}^k$ and write

$$\mathbf{A}^{n+k} = \left[- a_{n-1}\mathbf{A}^{n+(k-1)} + \ldots + a_{n-1}\mathbf{A}^{n+(k-k)} + a_{n-1}\mathbf{A}^{n-1} + \ldots + a_1\mathbf{A}^{k+1} + a_0\mathbf{A}^k\right] \tag{56}$$

$$k = 0, 1, \ldots$$

We can substitute from eq.(56) for all terms $\mathbf{A}^n$ and higher in eq.(54) and write the following equation with finite terms with no power of **A** higher than n–1.

$$\Phi(t) = e^{\mathbf{A}t} = \alpha_0(t)\mathbf{I} + \alpha_1(t)\mathbf{A} + \alpha_2(t)\mathbf{A}^2 + \alpha_3(t)\mathbf{A}^3 + \ldots + \alpha_{n-1}(t)\mathbf{A}^{n-1} \tag{57}$$

Under the assumption that the eigenvalues $\{\lambda_k, k = 1, 2, \ldots, n\}$ of **A** are distinct we have the following equation:

$$\begin{bmatrix} e^{\lambda_1 t} \\ e^{\lambda_2 t} \\ \vdots \\ e^{\lambda_n t} \end{bmatrix} = \begin{bmatrix} 1 & \lambda_1 & \lambda_1^2 \cdots \lambda_1^{n-1} \\ 1 & \lambda_2 & \lambda_2^2 \cdots \lambda_2^{n-1} \\ & & \vdots \\ 1 & \lambda_n & \lambda_n^2 \cdots \lambda_n^{n-1} \end{bmatrix} \cdot \begin{bmatrix} \alpha_0(t) \\ \alpha_1(t) \\ \vdots \\ \alpha_{n-1}(t) \end{bmatrix} \tag{58}$$

from which the coefficients $\{\alpha_k(t), k = 0, 1, \ldots n-1\}$ can be calculated. Having obtained the state transition matrix $\Phi(t)$, we can evaluate the state vector $\mathbf{x}(t)$ and the impulse response matrix $\mathbf{h}(t)$ by convolution. Thus,

$$\begin{aligned} \mathbf{x}(t) &= \int_0^t \Phi(t-\tau)\mathbf{B}\mathbf{u}(\tau)d\tau \\ \mathbf{h}(t) &= \int_0^t \Phi(t-\tau)\mathbf{B}\delta(\tau)d\tau = \Phi(t)\mathbf{B} \end{aligned} \tag{59}$$

Example

We shall now calculate the state transition matrix for $\mathbf{A}_1$ from canonical form 1. The characteristic equation is $s^4 + 10s^3 + 35s^2 + 40s + 24$ and the eigenvalues are $s_1 = -4$, $s_2 = -3$, $s_3 = -2$, $s_4 = -1$. From eq.(57) the state transition matrix $\Phi(t)$ satisfies,

$$\Phi(t) = e^{\mathbf{A}_1 t} = \alpha_0(t)\mathbf{I} + \alpha_1(t)\mathbf{A}_1 + \alpha_2(t)\mathbf{A}_1^2 + \alpha_3(t)\mathbf{A}_1^3 \tag{60}$$

and from eq.(58) we can write,

$$\begin{bmatrix} e^{-4t} \\ e^{-3t} \\ e^{-2t} \\ e^{-t} \end{bmatrix} = \begin{bmatrix} 1 & -4 & 16 & -64 \\ 1 & -3 & 9 & -27 \\ 1 & -2 & 4 & -8 \\ 1 & -1 & 1 & -1 \end{bmatrix} \begin{bmatrix} \alpha_0(t) \\ \alpha_1(t) \\ \alpha_2(t) \\ \alpha_3(t) \end{bmatrix} \tag{61}$$

The functions of time $\alpha_0(t)$, $\alpha_1(t)$, $\alpha_2(t)$, $\alpha_3(t)$ can be solved from eq.(61) and substituted into eq.(60) to obtain the state transition matrix $\Phi(t)$. The functions of time $\alpha_0(t)$, $\alpha_1(t)$, $\alpha_2(t)$, $\alpha_3(t)$ are obtained from

$$\begin{bmatrix} \alpha_0(t) \\ \alpha_1(t) \\ \alpha_2(t) \\ \alpha_3(t) \end{bmatrix} = \begin{bmatrix} -1 & 4 & -6 & 4 \\ -\dfrac{11}{6} & 7 & -\dfrac{19}{2} & \dfrac{13}{3} \\ -1 & \dfrac{7}{2} & -4 & \dfrac{3}{2} \\ -\dfrac{1}{6} & \dfrac{1}{2} & -\dfrac{1}{2} & \dfrac{1}{6} \end{bmatrix} \begin{bmatrix} e^{-4t} \\ e^{-3t} \\ e^{-2t} \\ e^{-t} \end{bmatrix}$$

$$= \begin{bmatrix} -e^{-4t} + 4e^{-3t} - 6e^{-2t} + 4e^{-t} \\ -\dfrac{11}{6}e^{-4t} + 7e^{-3t} - \dfrac{19}{2}e^{-2t} + \dfrac{13}{3}e^{-t} \\ -e^{-4t} + \dfrac{7}{2}e^{-3t} - 4e^{-2t} + \dfrac{3}{2}e^{-t} \\ -\dfrac{1}{6}e^{-4t} + \dfrac{1}{2}e^{-3t} - \dfrac{1}{2}e^{-2t} + \dfrac{1}{6}e^{-t} \end{bmatrix} \qquad (62)$$

Substituting eq.(62) into eq.(60) we can compute the state transition matrix $\Phi(t)$ for $\mathbf{A_1}$ for $t \geq 0$, that is given by:

$\Phi(t) =$

$$\begin{bmatrix} -e^{-4t} + 4e^{-3t} - 6e^{-2t} + 4e^{-t} \\ 4e^{-4t} - 12e^{-3t} + 126e^{-2t} - 4e^{-t} \\ -16e^{-4t} + 36e^{-3t} - 24e^{-2t} + 4e^{-t} \\ 64e^{-4t} - 108e^{-3t} + 48e^{-2t} - 4e^{-t} \end{bmatrix}$$

$$\begin{matrix} -\dfrac{11}{6}e^{-4t} + 7e^{-3t} - \dfrac{19}{2}e^{-2t} + \dfrac{13}{3}e^{-t} \\ \dfrac{22}{3}e^{-4t} - 21e^{-3t} + 19e^{-2t} - \dfrac{13}{3}e^{-t} \\ -\dfrac{88}{3}e^{-4t} + 63e^{-3t} - 38e^{-2t} + \dfrac{13}{3}e^{-t} \\ \dfrac{352}{3}e^{-4t} - 189e^{-3t} + 76e^{-2t} - \dfrac{13}{3}e^{-t} \end{matrix}$$

$$
\begin{matrix}
-e^{-4t} + \frac{7}{2}e^{-3t} - 4e^{-2t} + \frac{3}{2}e^{-t} \\
4e^{-4t} - \frac{21}{2}e^{-3t}t + 8e^{-2t} - \frac{3}{2}e^{-t} \\
-16e^{-4t} + \frac{63}{2}e^{-3t} - 16e^{-2t} + \frac{3}{2}e^{-t} \\
64e^{-4t} - \frac{189}{2}e^{-3t} + 32e^{-2t} - \frac{3}{2}e^{-t}
\end{matrix}
\qquad
\left.\begin{matrix}
-\frac{1}{6}e^{-4t} + \frac{1}{2}e^{-3t} - \frac{1}{2}e^{-2t} + \frac{1}{6}e^{-t} \\
\frac{2}{3}e^{-4t} - \frac{3}{2}e^{-3t} + e^{-2t} - \frac{1}{6}e^{-t} \\
-\frac{8}{3}e^{-4t} + \frac{9}{2}e^{-3t} - 2e^{-2t} + \frac{1}{6}e^{-t} \\
\frac{32}{3}e^{-4t} - \frac{27}{2}e^{-3t} + 4e^{-2t} - \frac{1}{6}e^{-t}
\end{matrix}\right] \quad (63)
$$

and the state impulse response matrix $\mathbf{h}(t)$ is given by,

$$
\mathbf{h}(t) = \Phi(t)\begin{bmatrix} 0 \\ 0 \\ 0 \\ 1 \end{bmatrix} = \begin{bmatrix} -\frac{1}{6} & \frac{1}{2} & -\frac{1}{2} & \frac{1}{6} \\ \frac{2}{3} & -\frac{3}{2} & 1 & -\frac{1}{6} \\ -\frac{8}{3} & \frac{9}{2} & -2 & \frac{1}{6} \\ \frac{32}{3} & -\frac{27}{2} & 4 & -\frac{1}{6} \end{bmatrix} . \begin{bmatrix} e^{-4t} \\ e^{-3t} \\ e^{-2t} \\ e^{-t} \end{bmatrix} u(t) \quad (64)
$$

The scalar output impulse response $h_o(t) = \mathbf{C}\mathbf{h}(t)$ where $\mathbf{C}$ is a row vector given by [30 11 1 0]. Or,

$$
\begin{aligned}
h_o(t) &= [\,30\ 11\ 1\ \ 0\,]\ \mathbf{h}(t) \\
&= -\frac{1}{3}e^{-4t} + 3e^{-3t} - 6e^{-2t} + \frac{10}{3}e^{-t} \quad t \geq 0
\end{aligned} \quad (65)
$$

As can be seen, solving for trajectories through the time domain route is tedious and time consuming. The easier method is to use the frequency domain technique discussed below.

Frequency Domain Method

The Laplace transforms for eq.(4) as shown in eqs.(40) are given below.

$$\begin{aligned}\mathbf{X}(s) &= \boldsymbol{\Phi}(s)\,\mathbf{BU}(s)\\ \mathbf{Y}(s) &= \mathbf{CX}(s) + \mathbf{DU}(s) = \left[\mathbf{C\Phi}(s)\mathbf{B} + \mathbf{D}\right]\mathbf{U}(s\end{aligned} \qquad (40)$$

We take partial fraction expansion of each individual term of $\mathbf{X}(s)$. The denominators of each of the terms of the $n \times n$ $\boldsymbol{\Phi}(s)$ are the characteristic polynomial of degree n having n distinct roots. We shall assume that each term of $\mathbf{U}(s)$ is of degree d_k, $k = 1, \ldots, r$, where d_k will be less than n for all k. In eq.(40) we can form the partial fractions for each of the terms $\boldsymbol{\Phi}(s)$, $\mathbf{U}(s)$, $\mathbf{X}(s)$, and $\mathbf{Y}(s)$. From eq.(41),

$$\boldsymbol{\Phi}(s) = \left[\phi_{.1}(s)\,\phi_{.2}(s) \;\cdots\; \phi_{.j}(s) \;\cdots\; \phi_{.n}(s)\right] \qquad (66)$$

and each of the vectors $\phi_{.j}$, $j = 1, \ldots, n$ can be expressed as partial fractions as follows.

$$\phi_j(s) = \left[\phi_{1j}(s)\;\; \phi_{2j}(s) \ldots \phi_{ij}(s) \ldots \phi_{nj}(s)\right]^T$$

$$= \begin{bmatrix} k_{1j}^1 & k_{1j}^2 & \cdots & k_{1j}^r & \cdots & k_{1j}^n \\ k_{2j}^1 & k_{2j}^2 & \cdots & k_{2j}^r & \cdots & k_{2j}^n \\ & & & \vdots & & \\ k_{ij}^1 & k_{ij}^2 & \cdots & k_{ij}^r & \cdots & k_{ij}^n \\ & & & \vdots & & \\ k_{nj}^1 & k_{nj}^2 & \cdots & k_{nj}^r & \cdots & k_{nj}^n \end{bmatrix} \cdot \begin{bmatrix} \dfrac{1}{s+s_1} \\ \dfrac{1}{s+s_2} \\ \vdots \\ \dfrac{1}{s+s_r} \\ \vdots \\ \dfrac{1}{s+s_n} \end{bmatrix} \qquad (67)$$

and the partial fraction expansion for $\mathbf{U}(s)$ is,

$$
\mathbf{U}(s) = \begin{bmatrix} \dfrac{\alpha_{11}}{s + s_{\alpha 11}} + \dfrac{\alpha_{12}}{s + s_{\alpha 12}} + \cdots + \dfrac{\alpha_{1d_1}}{s + s_{\alpha 1 d_1}} \\ \dfrac{\alpha_{21}}{s + s_{\alpha 21}} + \dfrac{\alpha_{22}}{s + s_{\alpha 22}} + \cdots + \dfrac{\alpha_{2d_2}}{s + s_{\alpha 2 d_2}} \\ \vdots \\ \dfrac{\alpha_{m1}}{s + s_{\alpha m1}} + \dfrac{\alpha_{m2}}{s + s_{\alpha m2}} + \cdots + \dfrac{\alpha_{md_r}}{s + s_{\alpha m d_r}} \end{bmatrix} \tag{68}
$$

From eq.(40a) $\mathbf{X}(s)$ can be given by,

$$
\mathbf{X}(s) = \Phi(s) \begin{bmatrix} b_{11} & b_{12} & \cdots & b_{1m} \\ b_{21} & b_{22} & \cdots & b_{2m} \\ \vdots & \vdots & \vdots & \vdots \\ b_{n1} & b_{n2} & \cdots & b_{nm} \end{bmatrix} \mathbf{U}(s) \tag{69}
$$

$$
\mathbf{Y}(s) = \begin{bmatrix} c_{11} & c_{12} & \cdots & c_{1n} \\ c_{21} & c_{22} & \cdots & c_{2n} \\ \vdots & \vdots & \vdots & \vdots \\ c_{r1} & c_{r2} & \cdots & c_{rn} \end{bmatrix} .\mathbf{X}(s) \tag{70}
$$

Each one of the terms in eqs.(68-70) is of the form $[s + s_p]^{-1}$ representing the time function $e^{-s_p t}$, with p ranging through all the indices of $\mathbf{U}(s)$, $\mathbf{X}(s)$ and $\mathbf{Y}(s)$. Equations (69, 70) may look formidable but in an actual problem they are quite simple.

Example

We shall find the trajectories using Laplace transform methods for the same example solved by the time domain techniques. The resolvent matrix $\Phi(s)$ given in eq.(44) can be expressed as a partial fraction. The partial fraction expansions for $\phi_{.j}(s)$, j = 1, 2, 3, 4 of eq.(67) are shown below.

$$\phi_{.1}(s) = \begin{bmatrix} -1 & 4 & -6 & 4 \\ 4 & -12 & 12 & -4 \\ -16 & 36 & -24 & 4 \\ 64 & -108 & 48 & -4 \end{bmatrix} \cdot \begin{bmatrix} \frac{1}{s+4} \\ \frac{1}{s+3} \\ \frac{1}{s+2} \\ \frac{1}{s+1} \end{bmatrix}$$

$$\phi_{.2}(s) = \begin{bmatrix} -\frac{11}{6} & 7 & -\frac{19}{2} & \frac{13}{3} \\ \frac{22}{3} & -21 & 19 & -\frac{13}{3} \\ -\frac{88}{3} & 63 & -38 & \frac{13}{3} \\ \frac{352}{3} & -189 & 76 & -\frac{13}{3} \end{bmatrix} \cdot \begin{bmatrix} \frac{1}{s+4} \\ \frac{1}{s+3} \\ \frac{1}{s+2} \\ \frac{1}{s+1} \end{bmatrix}$$

$$\phi_{.3}(s) = \begin{bmatrix} -1 & \frac{7}{2} & -4 & \frac{3}{2} \\ 4 & -\frac{21}{2} & 8 & -\frac{3}{2} \\ -16 & \frac{63}{2} & -16 & \frac{3}{2} \\ 64 & -\frac{189}{2} & 32 & -\frac{3}{2} \end{bmatrix} \cdot \begin{bmatrix} \frac{1}{s+4} \\ \frac{1}{s+3} \\ \frac{1}{s+2} \\ \frac{1}{s+1} \end{bmatrix}$$

$$\phi_{.4}(s) = \begin{bmatrix} -\frac{1}{6} & \frac{1}{2} & -\frac{1}{2} & \frac{1}{6} \\ \frac{2}{3} & -\frac{3}{2} & 1 & -\frac{1}{6} \\ -\frac{8}{3} & \frac{9}{2} & -2 & \frac{1}{6} \\ \frac{32}{3} & -\frac{27}{2} & 4 & -\frac{1}{6} \end{bmatrix} \cdot \begin{bmatrix} \frac{1}{s+4} \\ \frac{1}{s+3} \\ \frac{1}{s+2} \\ \frac{1}{s+1} \end{bmatrix} \qquad (71)$$

The state variable transform vector **X**(s) is from eq.(40a),

$$\mathbf{X}(s) = \left[\phi_{.1}(s)\,\phi_{.2}(s)\,\phi_{.3}(s)\,\phi_{.4}(s)\right]\cdot\begin{bmatrix}0\\0\\0\\1\end{bmatrix}.\mathbf{U}(s$$

$$= \phi_{.4}(s)\ \mathbf{U}(s)$$

$$= \begin{bmatrix} \frac{1}{6} & \frac{1}{2} & \frac{1}{2} & \frac{1}{6} \\ \frac{2}{3} & -\frac{3}{2} & 1 & -\frac{1}{6} \\ -\frac{8}{3} & \frac{9}{2} & -2 & \frac{1}{6} \\ \frac{32}{3} & -\frac{27}{2} & 4 & -\frac{1}{6} \end{bmatrix} \cdot \begin{bmatrix} \frac{1}{s+4} \\ \frac{1}{s+3} \\ \frac{1}{s+2} \\ \frac{1}{s+1} \end{bmatrix} .\mathbf{U}(s) \qquad (72)$$

Assuming that U(s) = 1 and taking the inverse Laplace transform of eq.(72), the trajectory vector **x**(t) under the assumption of zero initial conditions is the impulse response vector **h**(t),

$$\mathbf{x}(t) = \mathbf{h}(t) = \begin{bmatrix} -\frac{1}{6} & \frac{1}{2} & -\frac{1}{2} & \frac{1}{6} \\ \frac{2}{3} & -\frac{3}{2} & 1 & -\frac{1}{6} \\ -\frac{8}{3} & \frac{9}{2} & -2 & \frac{1}{6} \\ \frac{32}{3} & -\frac{27}{2} & 4 & -\frac{1}{6} \end{bmatrix} \cdot \begin{bmatrix} e^{-4t} \\ e^{-3t} \\ e^{-2t} \\ e^{-t} \end{bmatrix} u(t) \qquad (73)$$

and eq.(73) is the same as eq.(64). The corresponding y(t) is the output impulse response $h_o(t)$ given in eq.(65)

Complete State Variable Solution

The state transition matrix Φ(t) plays an important role in the general solution of the state equations given in eq.(4).

$$\frac{d\mathbf{x}(t)}{dt} = \mathbf{Ax}(t) + \mathbf{Bu}(t)$$

$$\mathbf{y}(t) = \mathbf{Cx}(t) + \mathbf{Du}(t) \qquad (4)$$

The general solution of the state equations can be obtained, given the initial condition matrix $x(t_{0+})$ and the state transition matrix $\Phi(t) = e^{At}$. We multiply both sides of the first equation in (4) by e^{-At} yielding,

$$e^{-At}.\frac{d\mathbf{x}(t)}{dt} = e^{-At}.\mathbf{Ax}(t) + e^{-At}.\mathbf{Bu}(t) \qquad (74)$$

Rearranging the terms in eq.(74) we obtain,

$$e^{-At}.\frac{d\mathbf{x}(t)}{dt} - e^{-At}.\mathbf{Ax}(t) = e^{-At}.\mathbf{Bu}(t)$$

$$\frac{d}{dt}\left[e^{-At}.\mathbf{x}(t)\right] = e^{-At}.\mathbf{Bu}(t) \qquad (75)$$

Integrating both sides of the second equation in (75) from t_{0+} to t we can write,

$$\int_{t0+}^{t} \frac{d}{dt}\left[e^{-A\tau}.\mathbf{x}(\tau)d\tau\right] = \int_{t0+}^{t} e^{-A\tau}.\mathbf{Bu}(\tau)d\tau$$

$$e^{-At}.\mathbf{x}(t) - e^{-A\,t0+}.\mathbf{x}(\,t_{0+}) = \int_{t0+}^{t} e^{-A\tau}.\mathbf{Bu}(\tau)d\tau \qquad (76)$$

Multiplying both sides of the second equation in (76) by e^{At} results in,

$$\mathbf{x}(t) = e^{\mathbf{A}(t-t0+)}.\mathbf{x}(\,t_{0+}) + \int_{t0+}^{t} e^{\mathbf{A}(t-\tau)}.\mathbf{Bu}(\tau)d\tau \qquad (77)$$

and the corresponding expression for y(t) is obtained from eq.(4) as,

$$\mathbf{y}(t) = \mathbf{C}e^{\mathbf{A}(t-t0+)}.\mathbf{x}(\,t_{0+}) + \int_{t0+}^{t} \mathbf{C}e^{\mathbf{A}(t-\tau)}.\mathbf{Bu}(\tau)d\tau + \mathbf{Du}(t)$$

Or,

$$\mathbf{y}(t) = \mathbf{C}e^{\mathbf{A}(t-t0+)}.\mathbf{x}(t_{0+}) + \int_{t0+}^{t} \left[\mathbf{C}e^{\mathbf{A}(t-\tau)}.\mathbf{B} + \mathbf{D}\delta(t-\tau)\right] \mathbf{u}(\tau)d\tau \quad (78)$$

Note that the expression given for **y**(t) in eq.(78) is only a representation and it is not very easy to compute **y**(t) using this equation.

Properties of State Transition Matrices (STM)

1. Derivative Property
   $\Phi(t, t_0)$ satisfies the matrix differential equation,
   $$\frac{\partial \Phi(t, t_0)}{\partial t} = A\Phi(t, t_0)$$
   with initial conditions $\Phi(t_0, t_0) = \mathbf{I}$.

2. Consistency Property
   $\Phi(t, t) = \mathbf{I}$ for all t in T

3. Semi-group Property (Transition)
   $\Phi(t_2, t_1)\Phi(t_1, t_0) = \Phi(t_2, t_0)$

4. Inversion Property
   $\Phi^{-1}(t, t_0) = \Phi(t_0, t)$

5. Separation Property
   $\Phi(t_1, t_0) = \Phi(t_1).\Phi^{-1}(t_0)$

REFERENCES

1. M. Abromowitz and I. A. Stegun, *Handbook of Mathematical Functions*, Dover Publications, 1972.
2. G. Arfken, *Mathematical Methods for Physicists*, Academic Press, 1985.
3. E. O. Brigham, *The Fast Fourier Transform*, Prentice-Hall, 1974.
4. E. O. Brigham, *The Fast Fourier Transform and its Applications,* Prentice-Hall, 1988.
5. P. K. Das, *Optical Signal Processing Fundamentals,* Springer-Verlag, 1991.
6. R. C. Dorf and R. J. Tallarida, *Pocketbook of Electrical Engineering Formulas*, CRC Press, 1993.
7. F. R. Gantmacher, *The Theory of Matrices*, Chelsea Publishing, 1959.
8. M. J. Greenberg, *Advanced Engineering Mathematics*, Prentice-Hall, 1988.
9. T. Kailath, *Linear Systems*, Prentice-Hall, 1980.
10. S. Haykin, *Adaptive Filter Theory*, Second Edition, Prentice-Hall, 1991.
11. M. O'Flynn and E. Moriarty, *Linear Systems, Time Domain and Transform Analysis*, John Wiley, 1987.
12. G. E. Shilov, *An Introduction to Theory of Linear Spaces*, Prentice-Hall, 1963.
13. S. S. Soliman and M. D. Srinath, *Continuous and Discrete Signals and Systems*, Prentice-Hall, 1990.
14. Staff of Education and Research Associates, M. Fogiel, Director, *Handbook of Mathematical, Scientific, and Engineering Formulas, Tables, Functions, Graphs, Transforms*, 1988.

15. G. Strang, *Linear Algebra and its Applications*, Third Edition, Harcourt, Brace and Jovanovich, 1988.

INDEX